INNOVATION

CROWN
BUSINESS
NEW YORK

INNOVATION

THE FIVE DISCIPLINES FOR
CREATING WHAT CUSTOMERS WANT

CURTIS R. CARLSON AND WILLIAM W. WILMOT

Library of Congress Cataloging-in-Publication Data

Carlson, Curtis Raymond.

Innovation: the five disciplines for creating what customers want / Curtis R. Carlson and William W. Wilmot—1st ed.

Includes bibliographical references and index.

1. Technological innovations—Management—Case studies. 2. Organizational effectiveness—Case studies. 3. Creative ability in business—Case studies. 4. Industrial management—Case studies. 5. New products—Case studies. I. Wilmot, William W. II. Title.

HD45.C37 2006

658.4'063—dc22   2006003861

ISBN 13: 978-0-307-33669-9

ISBN 10: 0-307-33669-7

Printed in the United States of America

Design by Nora Rosanky

10 9 8 7 6 5 4 3 2 1

First Edition

This book is dedicated to our colleagues at SRI International and at our subsidiary, the Sarnoff Corporation, and to our partners and customers around the world. Special gratitude goes to the following teammates whose ideas, human values, and commitment to creating the highest customer value made it possible:

NORMAN WINARSKY

HERMAN GYR

LEONARD POLIZZOTTO

LASZLO GYORFFY

contents

DISCIPLINE 3: INNOVATION CHAMPIONS

DISCIPLINE 4: INNOVATION TEAMS

DISCIPLINE 5: ORGANIZATIONAL ALIGNMENT

INNOVATION

why Listen to us?

Hardly a day passes without someone commenting that the ability to innovate has become a survival issue for you as an individual, the enterprise you work for, and the country you live in. There's no arguing that point, and in fact we would go further and say that there's no more important issue today.

Given that, why should you read us? What do we bring to the party? Our answer is twofold: first, the world-changing innovations that SRI International helped create over sixty years. And second, a process and way of thinking about innovation that is intimately connected to the way enterprises function in both the short and the long term.

Day in and day out, the chances are that you're using an SRI innovation.[1] Here are a few examples:

- the computer mouse and the personal computer interface you use at work and home.

- the Internet designations .com, .org, and .gov, which were developed as a way to organize domain names.

- mobile communication, where we sent the first transmissions over wireless and wired networks. We also received the first Arpanet log-on, which was the precursor to the Internet.

- the squarish numbers on the bottom of your checks that enable your bank to maintain your account balance correctly.

- the tracking system that enables the U.S. Postal Service to quickly get mail to you from the millions of pieces that people send around the country daily

- Halofantrine, an antimalarial drug to combat one of the world's most lethal killers—which is only one among the drugs we have helped discover or develop, some of which treat HIV and cancer

- the high-definition television in your living room, for which we received an Emmy Award (one of the ten that we have received),[2] the highest award in broadcasting. We've also won an Academy Award for our contributions to motion picture production.

We have created dozens of new companies and pioneered revolutionary new business opportunities. Nuance Communications, for example, is the industry leader in computer speech recognition.[3] When you talk to an automated operator at a company such as Charles Schwab, you are probably talking to a computer that includes technology from SRI.

Consider also Intuitive Surgical, the leading minimally invasive surgical company, which uses technology pioneered at SRI. Instead of performing heart surgery, for example, by opening up a person's chest cavity, Intuitive Surgical's da Vinci surgical system allows a doctor across the room—or the planet—to use three small probes placed through the chest wall to perform the operation, seeing and feeling as if she is operating directly on the patient. Most important, the patient's stay in the hospital is also shortened due to faster healing, and that makes both patients and insurance companies happy. The time people stay in the hospital drops from five or eight days to two or three, and the time spent away from work is cut from two months to about one.

Or consider Artificial Muscle, Inc. Imagine that you could create motors and actuators that had the virtues of human muscle—motors that were light, efficient, powerful, and configured in a wide array of sizes and shapes. These motors have the potential to replace many small electric motors currently in use in cars and consumer appliances. Or consider PacketHop, a business that potentially takes telecommunication companies out of the loop by enabling cell phones to receive and transmit messages by creating their own ad hoc network without going through a phone company.

At SRI International we have completed tens of thousands of projects and helped create hundreds of billions of dollars of marketplace value. We have served customers in more than half the countries in the world, and we work in all key technical areas—infotech, biotech, and nanotech. Today we are working to further revolutionize and improve our world, from producing clean energy, to transforming the way drugs are developed, to making the educational system much more effective, to securing the Internet and the U.S. homeland.

At SRI we are motivated to work on problems that are important, not just interesting. For example, we have demonstrated a fuel cell to burn dirty coal efficiently and cleanly, a system to produce titanium at almost the cost of aluminum, and a computer to teach freshman physics. We are also developing drugs to remediate the effects of addiction, and we are creating the next generation of intelligent, mobile communication devices.

As we will make clear, these great innovations were not done exclusively at SRI. Rather we had thousands of superb customers and partners who all did their part in transitioning the ideas, inventions, and innovations of SRI into the marketplace. We cannot thank them enough for their contributions.

As these examples suggest, we live in a *world of abundance.* Unlike the industrial age, improvements in knowledge-based products and services have no upper limits. The opportunities are effectively limitless if you can innovate. Innovation leads to prosperity and a higher quality of life. It is the basis for increased productivity, competitiveness, and national wealth. And ultimately, the major problems of our age—poverty, health, and the environment—will only be addressed through our collective ability to innovate. But an ability to innovate rapidly is required if we are to identify and develop the abundance that is before us.

That leads to the second thing we bring to the table: We have developed a disciplined process of innovation that is practically linked to the way things get done in an enterprise. This disciplined process of innovation is the centerpiece of the book—the Five Disciplines of *Innovation.*

Many people are confused about what leads to successful innovations. Innovation is not just the invention of some clever new gadget. It is much more than that. Innovation is the successful creation and

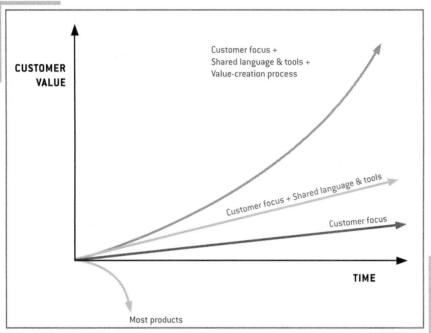

Figure 1.1: Enterprises that focus on their customers have shared language and tools for understanding customer value, and have a value-creation process do the best at creating new customer value.

delivery of a new or improved product or service in the marketplace. Or to put it another way, innovation is the process that turns an idea into value for the customer and results in sustainable profit for the enterprise. Innovations can be incremental (a child's new toy) or transformational (the development of instant photography). In all cases innovations deliver new customer value in the marketplace.

What we often find, though, is confusion about what innovation is and how it is accomplished. Instead of focusing on creating new customer value, we often find leadership in organizations trying to encourage "creativity" with the hope of achieving greater success. They sometimes even have large teams working to improve creativity throughout the enterprise. But the real question is, "Creativity for what?" Focusing just on creativity can lead to misplaced resources and frustration. For example, people often claim that creative organizations cannot be disciplined in their processes. We profoundly disagree. The most creative individuals, teams, and organizations are extremely disciplined. But it is a special kind of discipline—one that unleashes creativity in the service of developing important innovations.

Given the amount of misplaced effort we see in organizations, we are not surprised by the high failure rates of many products and the decreasing lifetimes of companies. But these failures point to an exciting imperative: even a small improvement in our ability to innovate can have a huge impact. And unlike the exploitation of natural resources, there are no limits to growth when it comes to innovation.

The ability to innovate can be significantly, if not profoundly, improved in most enterprises. Figure 1.1 suggests the advances possible using innovation best practices. The negatively curved line at the bottom of the figure indicates that 80 to 90 percent of new products and services fail after a year or so.[4] They fail not because of technology, resources, or a dozen other reasons. They fail primarily because customers didn't want them: the enterprise did not understand its customers' needs. As indicated, an enterprise that focuses on the customer does better at creating customer value. An enterprise that focuses on the customer and has shared language and tools for understanding customer value does better still. Finally, an enterprise that focuses on the customer, has shared language and tools for understanding customer value, and has a systematic value-creation process does the best. The Five Disciplines of Innovation describe how you can achieve this more productive state in your team and enterprise.

With the emergence of India and China as major economic powers, the ability to innovate is moving to center stage as a survival issue for most companies and a competitiveness issue for most nations. Economic growth in many countries—such as Japan, France, and Germany—has stagnated over the past decade. In addition, few companies have demonstrated the ability to innovate systematically. Even fewer individuals have the innovation skills needed to take control of their careers and make the impact they are capable of in our hypercompetitive world.

This period is one of the most exciting, opportunity-rich times in history. It is also fraught with disruption and change. The Five Disciplines of Innovation will help you, your teams, your enterprise, and ultimately your nation thrive in this remarkable period. It will give you a strategy and plan to master innovation and move forward positively in our world of abundance.

tHe esseNce of INNOVatION:
HOW fRaNk HIt a HOme RuN

> "INNOVATION IS NOW THE PRIMARY DRIVER OF
> GROWTH, PROSPERITY, AND QUALITY OF LIFE."[1]
> *Paul M. Romer, Stanford University*

If you ask a group of people the simple question "What is innovation?" it's likely you will get a wide range of responses, such as:

- a new *technological breakthrough,* such as the discovery of the transistor
- a new *invention,* such as a one-wheel scooter
- a new *business model,* such as a no-frills airline
- a new *production process,* such as a lower-cost way to make computers
- a new *creative design,* such as a sleek, sexy automobile

Innovation certainly requires creativity, invention, and often the other ingredients listed above. But none of these alone is an innovation.

Innovation is the process of creating and delivering new customer value in the marketplace.[2] Until this happens, you may have invented something but you still don't have an innovation. A new system of delivering products, such as selling books via the Internet, once established, is an innovation.

It provides new customer value because it allows consumers to conveniently select and buy a wide variety of books from home. Of course for an innovation to be viable, it must be affordable and also provide a sustainable profit to the company.

Many people have invented revolutionary new products but have been unsuccessful at getting them into the marketplace. Philo Farnsworth invented television in 1927, but it was David Sarnoff who created television broadcasting to bring black-and-white television to the consumer in 1939.[3] He developed a successful business model that put together televisions, cameras, broadcasting stations, program content, and advertising. Farnsworth invented a device, while Sarnoff was the innovator who put all the pieces together to create an industry.

Sometimes the new customer value of an innovation may seem small, such as a better way for a nurse to administer a medical procedure or a better way to keep notes, such as 3M's Post-it Notes. At other times an innovation has revolutionary, worldwide impact, such as Wal-Mart's approach to mass product distribution and sales. Regardless of the magnitude of impact, all innovations provide additional customer value, with customer value defined as a product's or service's benefits minus its costs. Customer value includes not just tangible product or service benefits, such as that provided by a higher-resolution computer screen, but also intangible needs, such as service, convenience, and personal identity.

Consider convenience. The Apple iPod Nano can contain thousands of your favorite songs and can be attached to your key chain, allowing you to carry your music conveniently in your pocket. By addressing multiple aspects of customer value—product, convenience, service, experience, and emotion—companies such as Apple, Disney, Lexus, and Starbucks create significant customer value and can thus charge a premium.

The phrase *customer value* is thrown around so often that it may sound like just another buzz phrase. Yet the failure to think through the process of creating successful innovations by providing value to the customer is one reason that so many talented—and sometimes brilliant—people experience so much frustration. They can't find a way to be successful because they don't understand all the pieces that must be in place. This also explains why people in management often fail to act on potentially important new products and services. In the minds of many managers, far too many employees with allegedly innovative ideas are seen as a bother. They feel that employees don't understand the practical realities of the company's business and can't demonstrate to management why customers would care enough to buy the supposedly wonderful new widget. These managers have no process in place to engage employees and turn them from bothers into superstars of innovation.

That was certainly the case with Frank Guarnieri, a brilliant researcher who was full of new ideas but felt that he was banging his head against the wall. At this time I* was vice president in charge of new ventures when Frank visited me in my office. Frank hadn't been with us very long and we had never really gotten to know each other. His meeting with me was a final test to see if he could get some help with his idea. With dark hair and eyes radiating intelligence, Frank spoke with an unusual accent, a combination of "Brooklynese" and some obscure form of erudition, pronouncing words with unique emphasis. He exuded passion and sincerity, and I immediately felt that he was someone to listen to.

Frank asked, "Do you know anything about biology?" When I told him no, he rolled his eyes slightly. Frank was probably thinking that this meeting was going to be another waste of time. He had come to see me because I was his last stop before walking out the door. As I learned later, that was supposed to happen just after he saw me. I was the last box to be checked on the way out.

Not knowing any of this, I asked Frank what he was up to. He told me he had an idea and that he was having a hard time getting anyone interested in it. I asked him what it was. He said, "Do you know that proteins are huge molecules made up of many tens of thousands of atoms?"

"Yes," I said.

"Do you know why?" he asked.

I said, "No," and again I saw a slight eye roll.

He told me that proteins had to be huge because they had a small spot, called a binding site, where they did their work when interacting with other proteins. All those atoms in the molecule were needed because the binding site had to be very specific in what it did. Otherwise,

*This section is in the voice of coauthor Curtis R. Carlson when he was at SRI's subsidiary, the Sarnoff Corporation.

proteins would incorrectly interact with one another and our bodies wouldn't function. "That makes sense," I said, not fully understanding the implications of what he was after.

Frank went on to say that if something goes wrong with one of your proteins, you have an illness, and you need a drug to fix the defective protein. The problem is that drugs are tiny molecules compared with proteins. They must attach themselves to exactly the right place on a particular protein to do their job. And to make this harder, if the drug attaches itself to any other protein in your body, it may cause a side effect. These two criteria make it incredibly hard to develop drugs. You must fix the broken protein and not interact with any others. Drug development is like walking into a football arena with 100,000 unknown spectators and trying to find the one person who will best connect with you and no one else. This is why a new drug can take a billion dollars and fifteen years to develop. It is truly like finding a needle in a haystack.

This was fascinating, so I asked, "What's your idea?"

He said, "By calculation in a computer I can design small molecules that will attach themselves to just the right place on a protein to fix the problem."

"Wow," I thought, "this is a *very* important problem." On my checklist of whether we should support a project, Frank had just hit the jackpot. If he could do this, it would be a huge contribution to medicine. The question was "Could Frank really do this?" What he was describing was a very difficult calculation—maybe impossible.

I asked Frank to explain his approach, but we didn't get very far. Like many brilliant people, Frank assumed others knew as much as he did. I certainly didn't. And as is often true at this point, his idea raised more questions than it answered. For example, I asked Frank about competing approaches. He didn't know of any direct competition, although I knew from past experience that there is always competition.

At this point I had to go to a meeting. Frank asked, "What should I do next, write up a hundred-page report and come back in a month?"

"No," I said. "Why don't you come back at seven tonight and bring answers to the four questions that define your value proposition." I wanted Frank to know that I would support him by acting quickly too.

"What are the four questions?" Frank asked.

I said, "What is the market *need*? What is *your approach* to addressing this need? What are the *benefits per costs* of your approach? And how do those benefits per costs compare with the *competition*?" I told him we referred to these four questions as a "value proposition" or, in short, "NABC" for *need, approach, benefits per costs, and competition. "They are the fundamentals; it doesn't make sense to write up a big report until we can explain them in simple language to a knowledgeable person."

Frank said, "Okay."

Before Frank left I asked him one last question. I wanted to know, if I became his partner and worked with him to develop his idea, whether he would agree to be the project *champion* and do the work necessary to make his idea a success.

Frank asked, "What does it mean to be a champion?"

I went through the standard catechism of commitment, teamwork, corporate responsibility, engagement in the process to create a compelling value proposition, and perseverance.

Frank said, with eyebrows raised and some skepticism, "Really. What is the probability of success?"

I answered, "If you will do these things, the probability, in my experience, is very high—in some situations close to one hundred percent." Frank still looked a little doubtful, so I said, "I understand that this is a lot to sign up for on day one, so why don't we get started. But I will ask you again in a week or two if you are the champion, and if you say no, that is the end of the project."

He said, "I understand," and we started.

Frank created his first value proposition and brought it back that evening. As expected, his initial attempt was confusing and incomplete. They always are. Spelling out an idea so that a busy person "gets it" quickly is one of the toughest challenges in the early stages of a new innovation, no matter how incremental. And this was clearly not an incremental innovation. It takes dozens of revisions—or iterations—to make a value proposition clear. But it was a start, we had begun, and Frank was on to something. He came back again and again, and each time his ideas were expressed a little more clearly. Frank introduced me to another terrific colleague, John Kulp. This was a big step forward. Now we had two people who had real knowledge in the field—plus me. I contributed knowledge of the disciplined process we had to go through and what the answers to our four questions needed to look like when we were done.

Over time we gathered other critical colleagues.[4] Frank talked to hundreds of people, interacted with potential clients and strategic partners, and constantly improved and refined his NABC value proposition. It was quite an experience: Every "expert" in the field told Frank that his ideas were impossible. They ignored him, rejected him, and sometimes insulted him. But Frank was always respectful and persevered.

I never had to ask Frank again if he was the champion. His passion took over and, after his first value proposition pitch, he was on his way. He didn't quit.

It took Frank and his team over a year of challenging work. When we started the process, I had asked him what his goal was. He said, "My goal is a research project with funding of a few hundred thousand dollars a year, but I would like to do much better than that."

I said, "I think we can." Frank eventually became a founding member in an exciting new company, Locus Pharmaceuticals,[5] with a multimillion-dollar research program and with some of the best people in his field. Just as important, he learned many of the skills and the disciplines needed to create successful innovations over and over again.

Since its formation Locus has made remarkable progress. At one point Frank and his colleagues asked for $40 million in new capital and they eventually decided to take $80 million—this at a time when most companies were having trouble attracting $10 million in venture capital. The reason for the excitement was that Locus was doing something important that had never been done, designing a new drug completely in a computer. This development has the potential of removing many tens of millions of dollars from the cost of drug development while reducing the development time by years—a transformational innovation.

Locus now has several potential blockbuster drugs that are close to clinical trials in cancer and inflammation—two vitally important areas of medicine. These amazing achievements were set in motion with the many iterations of Frank's first value proposition and a *discipline* of innovation that resulted in the creation of Locus Pharmaceuticals. Although Locus still has a long way to go before their drugs are proven to be safe and effective, a critical milestone has been reached in the history of drug development. Frank and his teammates hit a home run.

When I became CEO of SRI International in 1998, I joined an organization with an impressive history of innovation—an organization renowned for creating world-changing products and services. SRI's many achievements have helped form new industries. In addition to the list of achievements given in the preface, SRI has made many contributions across a wide spectrum of societal needs: From medical ultrasound imaging, to the foundations of credit cards, to over-the-horizon radar, to economic development in dozens of countries around the world (we received the Royal Order of the North Star from the King of Sweden for economic developments in Sweden), to programs to help disabled children learn better, to innovations such as the 911 call system, the term "stakeholder," and the fundamental business concept of SWOT (strengths, weaknesses, opportunities, and threats).

Despite this enviable history of achievement, SRI had stopped growing in the 1990s. The *discipline* of innovation that provided Frank and his organization with the skills to be successful and thrive in today's knowledge world wasn't part of the organization's DNA. As was the case at SRI and in organizations around the world, the process of innovation being used was obsolete. Here was an extraordinary organization whose researchers had produced many life-improving innovations, and yet it was losing ground. In our fast-paced age, if SRI wanted to thrive, it would have to create higher-value innovations more rapidly. But how?

How We Got Here

In analyzing the fundamentals of successful marketplace innovation, we discovered that much had been written about either the "lone genius in a garage" or the "eureka!" moment resulting from the "miracle" of invention. What was lacking was an explanation of *how* we could innovate successfully to create compelling customer value. Hoping for success from a lone genius or a miracle is not a realistic plan for organizational survival. If the genius decides to walk out the door one day and not come back, then what?

We needed a way to make innovation a *discipline.* That is, innovation needed to be treated as a subject that could be systematically understood and taught, step by step, first to individuals, then to their teams, and ultimately become part of the DNA of the entire enterprise. We were motivated by a quote given to every employee at Toyota: "Whenever something—anything—is to be produced, there must be rules, or a systematized method of producing it. Whether or not the people who do the actual production fully understand those rules, that system has a deciding effect on product quality, cost, safety, and all essential determinants of success or failure."[6] This was our goal for Innovation, to create "a systematized method of producing it" based on fundamental innovation best practices—practices that could be taught and understood by all in an enterprise.

Other approaches we learned about seemed misdirected. People, for example, often focus on teamwork, creativity, and culture as the keys to organizational success. But teamwork, creativity, and culture are not business objectives. When an organization, for example, sends its executives down a river in a raft to learn teamwork or has them build brightly colored paper airplanes to learn creativity, something is profoundly wrong.[7] And when a change agent comes into an organization announcing a plan to redo everyone's "culture," is it surprising that the organization takes offense and resists? We felt a need to reframe creativity, teamwork, and culture so that they focused specifically on the goal of customer value creation and innovation across the enterprise.

One often hears CEOs talking about "shareholder value." This is not very helpful either. Yes, shareholder value is important, but talking about it doesn't align the enterprise and tell each employee what to do. It puts the cart before the horse. You can create great shareholder value only by focusing the entire organization on creating compelling *customer* value.

During this journey to systematize innovation—to make it a discipline—we partnered with organizations around the world, always seeking out those who had best practices. We worked with some of the leading venture capitalists, such as Mayfield Fund, U.S. Venture Partners, and Morgenthaler Ventures. We met with leading companies and foundations, such as Sony, Ritz Carlton, Swisscom, Motorola, Intel, IBM, Toyota, IDEO, Thomson, Wipro, and the Gates Foundation. We worked with numerous government agencies, such as the Defense Advanced Research Projects Agency (DARPA), the National Institutes of Health (NIH), the Department of Energy (DoE), and the National Science Foundation (NSF). We continually combined the best practices we discovered with new ones we invented, as the employees at SRI International tested these ideas every day. Our staff and partners told us what worked and what didn't, and we refined our ideas, again and again.

We discovered that what was missing was an integrated approach to the process of innovation. All the different elements needed to be brought together in the service of making high-value innovations virtually inevitable. We required a blueprint for the "how" of innovation and not just a description of the "what." We discovered that success required mastery of Five Disciplines. Ignoring or being unskilled at any one leads to failure.

The result of our work on innovation best practices is the ideas that are the foundation of this book you are now reading, the Five Disciplines of *Innovation*. We found that to be successful at innovation, you do not have to totally change your enterprise, fire all your people, or import arcane practices that no one understands. Instead of hoping for a one-time "miracle," you can use a series of practices, much as Frank did, that lead to successful value creation. The Five Disciplines of Innovation give everyone in the enterprise—from brilliant people like Frank to people who have become bored corporate and government time-servers—the opportunity to take control of their careers. For managers who have to make the decisions and provide the financing for new projects, it provides the missing link to the business fundamentals: who's the customer and competition, what's the value of the innovation and the business model, and how are we going to make it happen? It is ultimately about creating an enterprise that can thrive—one that is full of Franks at every level.

Our perspective is very much in support of the rank-and-file employee. The vast majority of employees want to do a good job and contribute to the success of their enterprise. However, much of the failure we see is because employees do not have the value-creation skills required to fully contribute. Although the Five Disciplines of Innovation can be applied at any level of the enterprise, it is ultimately the responsibility of senior management to champion these ideas throughout the entire enterprise, as will be described in Chapter 16. This is when the full impact is felt by customers, the enterprise, and employees.

These practices have been tested internally at SRI and shared with external clients enrolled in the SRI Discipline of Innovation workshop. SRI propelled itself into double-digit growth by using these ideas, at a time when the economy was lumbering along at single-digit rates. The processes practiced at SRI and taught in our workshop apply to all enterprises—companies, government agencies, nonprofits, and universities—because they all have customers and all can create greater customer value. Our success has prompted leading organizations from around the world to take our workshop, including the world's foremost broadcaster, the British Broadcasting Corporation (BBC); the huge Japanese telecommunications group Nippon Telephone and Telegraph (NTT); and the leading Taiwanese research organization, the Industrial Technology Research Institute (ITRI). Recently, the U.S. Department of Homeland Security's cyber team has taken the SRI workshop, as have several universities. These and other organizations are incorporating the Five Disciplines of Innovation within their enterprises and enhancing their ability to create new customer value.

In any organization, talk about introducing a new process always raises fears, because employees are concerned about how it will impact their work and career. The management of an international corporation, which had been working with us for three years, challenged us to convince its most difficult cynics. If we could get them on board, the company would apply our ideas across the entire organization. Sure enough, when the group of fifteen cynics arrived, they were *very* skeptical. They participated in the workshop, however, and learned to use the key skills we demonstrated. At the end of the workshop, they held a celebration for us in honor of "a much more productive and enjoyable way to work." These "cynics" were actually individuals who cared deeply about their organization and were dedicated to making a difference. They simply needed a map for the road ahead. The Five Disciplines of Innovation marked that route for them.

We are interested in improving organizational success throughout the full range of innovations, from incremental to transformational. For existing products and services, incremental innovations—through new features, lower costs, improved designs, and enhanced experiences—are the core of an ongoing business's activity. Transformational or "disruptive" innovations are major new market opportunities that come through breakthrough technologies and business models. As we will demonstrate, the Five Disciplines of Innovation not only allow you to make incremental improvements to existing products and services, but also provide the launching pad for creating wholly new products and services.

Economic developments are increasingly being driven by powerful forces, including globalization and the accelerating speed of innovation. Everyone, from people running huge companies to the individual consumer, experiences the global economy on a daily basis. Just go to Wal-Mart and look on the bottoms of products to see where they are made. They are produced throughout the world. The emergence of India in software and China in manufacturing offers two examples of globalization that are affecting your life every day. This world economy is full of opportunities, but it is moving quickly, and innovation is the only way to stay ahead of these fast-moving developments and increasing competitive pressures. They mandate rethinking your innovation processes, the behavior of your teams, and the structure of your enterprise. Your innovation practices must, of necessity, exceed those of your competition if you are to thrive. The development of compelling *new customer value per time per dollars invested* is a vital index of an organization's potential for survival.

Business as Usual Won't Work

Both the tempo of change and the opportunities for innovation will continue to increase. Change will accelerate, both for individuals and for companies. Whereas a generation ago lifetime employment at one company was the career path for many people, it is now the rare person who does the same job in the same company for fifteen to twenty-five years. Project managers, state employees, members of mom-and-pop operations, nonprofit staff, and people in high-tech companies are all experiencing a sense of disquiet. We can all feel the tectonic shifts occurring below the surface, and these shifts impact careers.

"Business as usual" is a recipe for disaster. Well-known companies are disappearing at extraordinary rates. Major players, such as Digital Equipment Corporation and Compaq in computing, Arthur Andersen in auditing and accounting, AT&T in communications, and RCA and Polaroid in consumer electronics, no longer exist. And while some flagship companies like GE, Johnson & Johnson, and Procter & Gamble are doing fine, it is shocking to see other huge companies like United Airlines, GM, Sears, AT&T, Ford, and HP all struggling to keep their heads above water. Even a juggernaut like Microsoft is experiencing a slower rate of growth, suggesting that it too will need to innovate more effectively to stay on top.[8] Bill Gates knows this. He once said, "We are just two years away from failure." By this he means that Microsoft needs to create the next wave of innovations every two years.

Our jobs are becoming more like those in Hollywood, where a team is formed to do a project and then the people move on to the next one. But this Hollywood model, while attractive in concept to some, is difficult to emulate and execute. In addition, all workers worry that their jobs could be in jeopardy overnight through outsourcing, offshoring, mergers, or company failure. A different, more positive approach is needed.

Traditional professional training is not enough if you are to adapt and thrive in this tumultuous business world; you must also have new innovation skills. If you know how to create customer value, regardless of your particular enterprise, you have a much greater chance to succeed and remain employable over your full working years. Otherwise, you may become obsolete.

It doesn't matter if you have an advanced degree and work in aerospace or are trained as a financial analyst and work in an insurance company; your expertise has to adapt to the world. If not you may, as the *New York Times* recently reported,[9] be like Saul, who has a Ph.D. in physics and drives a cab in New York City. Instead of always going deeper and deeper into your field, specializing even more, you must also learn how to create customer value.

Just asking, "Who are my customers and what value do I bring to them?" is how you begin. *Innovation* will guide you through the skill sets you need to compete. We believe that *all employees* must and can innovate for their customers and organizations to add more value. The Five Disciplines of Innovation apply first to you and your teams. But the biggest impact is realized when they are applied across the enterprise, and the organization's processes and culture become dedicated to the rapid creation of new customer value based on innovation best practices. We call this level of achievement Continuous Value Creation (CVC).

Before Henry Ford made mass production a reality at the beginning of the twentieth century, cost was a primary barrier for the automobile consumer. Cars were too expensive for middle-income and working-class people. For a long time after mass production was established, people still accepted relatively poor quality as an inevitable outcome of the manufacturing process. Then W. Edwards Deming developed a process to continually improve product quality and revolutionized the manufacturing world.[10] Deming, Taiichi Ohno at Toyota, and others tackled the quality problem, first in Japan and then worldwide. Today almost all manufacturing companies use a version of Deming's and Ohno's lean-manufacturing ideas.

But looking at a job or a company from a quality and cost perspective alone does not fully equip us to compete. We must make three additions. The first is to focus the entire enterprise on all aspects of customer value, not just quality and cost. The second is to improve our innovation processes so that the creation of high-value innovations is more reliable, less costly, and faster. The third is to make continuous improvements to today's products and services while capturing major new, breakthrough opportunities for the future.

Continuous Value Creation (CVC) is a natural extension of the work of Ford and Deming as we move from the industrial age to the knowledge age. "Continuous" means that all aspects of the enterprise are focused on creating the highest customer value. "Value Creation" means a process can be learned and applied throughout the enterprise.

Customer value has many dimensions: physical features, quality and durability, service and convenience, experience and trust, emotional appeal and identity, and costs. We intuitively understand what value means: it is the sum of a product's benefits when compared with its costs. For example, the quality of a McDonald's hamburger may not be high, but the cost is very low, which provides good customer value. Alternatively, if you are taking someone out to dinner at a $60-a-person restaurant, the food, service, and ambience must be superb to justify the expense. Value can be increased either by increasing benefits to the consumer or by lowering costs.

Increasingly, the customer *experience* is as important as the physical product. Consider the example of a seemingly small innovation—an improved medical procedure. For the customer this improvement can have enormous value if it reduces pain and anxiety. Anyone who has been in a hospital where an intravenous needle has not been properly inserted will understand the value of a good procedure. Similarly, sincere reassurance from a nurse or surgeon before an operation provides great customer value. These simple human gestures also reduce medical costs. Evidence has been building that doctors who treat their patients with respect experience fewer malpractice lawsuits than those who are less solicitous of their patients.[11] It has become apparent that our emerging economy provides vast opportunities for innovation based on the customer's experience. For example, the uncommon coffee experience Starbucks delivers to its customers partly explains why people are willing to pay two to three dollars for a cup of coffee when far cheaper alternatives are available elsewhere.

As globalization and rapid innovation become more pervasive, the need to improve innovation practices will go from a smoldering issue to a burning one. Sustained innovation centered on customer value is the only approach that works and the only way to meet future challenges. Even a small improvement in our ability to successfully create new products, services, and companies can have a significant positive impact on the economies of every country.

If you focus on customer value as the starting point of successful innovation, a series of important questions arise:

- Who is your customer?

- What is the customer value you provide and how do you measure it?

- What innovation best practices do you use to rapidly, efficiently, and systematically create new customer value?

This book will provide you with specific concepts and tools to answer these important questions based on the Five Disciplines of Innovation.

- **Important Needs:** Work on important customer and market needs, not just what is interesting to you (Chapters 3 and 4).

- **Value Creation:** Use the tools of value creation to create customer value fast (Chapters 5, 6, 7, 8, and 9, and the Appendix).

- **Innovation Champions:** Be an innovation champion to drive the value-creation process (Chapter 10).

- **Innovation Teams:** Use a multidisciplinary, team-based approach to innovation to create a collective, genius-level IQ (Chapters 11, 12, 13, and 14).

- **Organizational Alignment:** Get your team and enterprise aligned to systematically produce high-value innovations (Chapters 15 and 16).

Failure to ultimately satisfy any one of the Five Disciplines of Innovation will *probably lead to failure* on the project. Of course, in the beginning of a new innovation and as you go along, one or more of the Five Disciplines will be missing. The key is to recognize this quickly and then address the problem. Even a small improvement in each of the Five Disciplines can significantly improve the probability of success.

Since the Five Disciplines must all be satisfied to achieve success, they are multiplicative. This means that success equals important needs × value creation × innovation champions × innovation teams × organizational alignment. If you have a zero for any of the Five Disciplines, the chances of success go to zero as well. If you evaluate your team, for example, and decide that that each discipline is 0.5 satisfied, this means that the overall probability of success is only about 3 percent. However, if you raise the probability from 50 to 60 percent, the probability for success goes up by almost three times. This simple example illustrates one of the main points of this book, that by being more thoughtful and effective about each of the Five Disciplines of Innovation, you can significantly increase your ability to succeed.

The end customer takes center stage in our world. At the same time there are other "customers" who must be satisfied for an enterprise to thrive. Value must be produced for the company, the shareholders, the employees, and the public. It is a primary function of management to keep these stakeholders all in proper balance. But the starting point is always our customers and our ability to create value for them. The ability to rapidly create new value for customers will help your organization thrive and allow you to become more valuable too, just as Frank did. Everyone has ideas, but as Frank's story shows, the path to success means connecting those ideas with customers who can use them. Thus, the Five Disciplines of Innovation complete the linkages between customers, individual inventors, and management.

The creation of significant new customer value and speed of innovation are the critical measures for success. The rapidly expanding world economy provides the opportunities; the global competition provides the imperatives. The competitiveness of teams, companies, and nations depends on the ability to innovate.

The Five Disciplines of Innovation are in response to the economic forces of our times. The major driver of innovation is the rapid improvement in knowledge-based activities, such as is occurring in computers, communications, and many other industries. This improvement is so rapid and profound in its impact that we refer to it as the "exponential economy," as described in the next chapter.

CHAPTER 2

INNOvate OR DIe:
the exponential economy

> "THE EXPONENTIAL ECONOMY REQUIRES
> EXPONENTIAL IMPROVEMENT PROCESSES."[1]
> *Curtis R. Carlson*

> "THE MOST POWERFUL FORCE IN
> THE UNIVERSE IS COMPOUND INTEREST."[2]
> *Albert Einstein*

The World Is a Dot

Nadia answers the phone and in her charming English accent says, "Hello, how can I help you?" Trying to quickly figure out where she is—the United States, Ireland, or India—you ask and she says she's in Bombay, India, in a room with five hundred other call-service operators. She is knowledgeable, respectful, and pleasant. You are impressed with the service you get. After you hang up, you reflect on this simple transaction and the great force on display: globalization as powered by worldwide communication. If you have visited India or China, you know that their infrastructures are relatively primitive by Western standards. But that matters less in the knowledge age. To be successful in our world, it is often enough to have well-educated people, electricity, and a connection to the Internet.

The impact of globalization has just begun. One indication of this conclusion comes from Infosys, a leading Indian knowledge-management company with 50,000 employees and an average age of twenty-six, that recently placed advertisements announcing ten thousand new job openings. It received more than a million applications,[3] an indication of the hunger and drive of the one billion people in what is, primarily, a developing country. Imagine being able to select only one out of every hundred applicants for your enterprise. It is hard to appreciate what it means to have a billion people in your country. And, of course, China has another billion. These are smart, ambitious, hardworking, entrepreneurial people.

As we are starting to understand, many Indian and Chinese workers are extremely well educated. The Indian Institute of Technology, formed in the late 1950s and then helped along by President Kennedy and others, is now arguably the most competitive undergraduate science and engineering school in the world. Only 2.3 percent of applicants are accepted, making it four to five times more difficult to get into than Harvard or MIT.[4] In the United States, science and engineering are in decline in academia, even though they are essential building blocks for the future and the opportunities in these fields have never been better. For example, computer science Ph.D.s have dropped from a peak of 1,100 per year in 1992, although the market need has increased by more than 5 percent per year.[5] By way of contrast, in India there are three preferred educational paths: medicine, science, and engineering—not law, social sciences, and liberal arts. We once spoke with an engineer from Bangalore about how he got started. He said, "I wanted to be a lawyer but my family said they would disown me if I became one. They weren't kidding." Unlike the United States, India and China are creating the human capital to compete.[6]

America is of course well known for its adaptive, entrepreneurial culture. The Indians and Chinese, though, are working hard to surpass the United States and the rest of the world. They are full of ideas, enthusiasm, and energy. They see boundless opportunity, and they work day and night to create new businesses.

Much of Western Europe is different, with many people living in a world of scarcity. In France unemployment is more than 22 percent for males under twenty-five years of age.[7] If you are an uneducated minority male, the situation is much worse at 50 percent. The French government mandated a 35-hour workweek, in part to share what they believe are the limited jobs available. This approach, based on a scarcity model, only makes the situation worse because it is focused on carving up the available opportunities, not creating new ones.

We meet executives and government officials from many countries. On a recent trip to Europe we spoke to high-level government officials in Germany about opening a business that would create new, high-paying jobs there. During our discussion we were curious about why Germany had been losing roughly a thousand jobs a day for a decade, and how they planned to reverse this trend. One of the officials at first hesitated and then said matter-of-factly, "With good engineering." That was it. If he was joking, he didn't smile. There was no alternative proposal to fix this predicament, and they seemed uninterested in moving forward with new approaches. The problem isn't one of just using engineering skills to fine-tune existing operations. It takes cultural resolve, critical infrastructure, and supportive governmental policies to create an environment that will systematically result in major new opportunities based on innovation. Reflecting on our enthusiastic partners in Bangalore and Beijing, who work sixty to seventy hours a week, we thought, "This is distressing; many Europeans don't understand the consequences that will result from these global developments. The increased levels of competition we are seeing are just beginning."

Who is impacted by these changes? In addition to manufactured products, we are seeing these global developments in call centers and other easily outsourced jobs, such as basic legal tasks, software applications, and the reading of medical X-rays.[8] But organizations in India and China are rapidly moving up the value chain to include research and all forms of innovation. Many major international companies—including GE, HP, IBM, and Intel[9]—now have facilities in Bangalore and other sites around the world. They must, because the talent is superb and the costs are low. In addition, speed of development is critical to success. It allows work to be done almost twenty-four hours a day. Tasks are passed from one country to another as the sun moves around the globe. A new vocabulary is emerging to describe these developments. People talk about "virtual" or "F-to-F," where virtual means online and F-to-F means face-to-face. High-speed communication allows people across the globe to harness the genius of others. But getting people around the world to collaborate doesn't happen easily. For example, a current ironic phrase is "continuous multiple attention." What percentage of the time are people really participating at the end of a phone or video connection? Virtual team members are probably reading e-mail, surfing the Web, and working on other presentations, all at the same time.

Although there are difficulties in working around the world, these difficulties will be addressed as communication and collaboration tools improve. We have moved to a world of global innovation. Work must be done where it creates the best value. Companies as diverse as Wal-Mart and Dell are now globally integrated enterprises. Much of their success is achieved through business-model innovations based on leveraging these worldwide developments. Trying to stop a country from adopting these developments, or trying to isolate it economically, is like trying to prevent water from flowing downstream. Dams, after all, only work up to a point.

In an important best-selling book, Thomas Friedman argued that the world is now "flat" in the sense that it is becoming an even playing field.[10] This is increasingly true, but in another sense our world is not just flat, it has been reduced to the size of a dot. Information and capital move around the world at the speed of light. Location does not matter for many knowledge-intensive activities. The idea of business clusters as a family of neighboring activities that result in a local competitive advantage is being redefined. The full implications of this dispersal of activities around the world are yet to be fully understood.

Exponential Economy

The impact of globalization is getting the world's attention, but there is a larger, more important development emerging: the "exponential economy." It includes those increasingly large segments of the economy improving in price performance at rapid, exponential rates, including computers, communications, biotechnology, and consumer products.

Consider, as one familiar example of the exponential economy, computers. Gordon Moore, the cofounder of Intel, noted that computers have increased in price performance by roughly 100 percent every eighteen months for more than a hundred years.[11] Although "Moore's Law" is an empirical observation (not a "law"), computers have approximated Moore's Law since the beginning of the twentieth century while the technologies used have changed many times, from wooden mechanical structures, to relays, to vacuum tubes, to transistors, and now to integrated circuits.

Moore's Law is likely to continue well into the twenty-first century with the introduction of new technologies. Predictions are that by 2025 your desktop computer may have the raw computing speed of the human brain.[12] This is a capability we now almost have in supercomputers, as announced in 2005 by IBM.[13] If we can keep on improving at today's rates, by 2040 to 2060 our desktop computers may be faster than the collective computing power of the world's population.[14] Ray Kurzweil calls this the "Singularity" to note its significance.[15] Although raw speed does not automatically give you performance superior to humans,[16] we wryly note that a PC with these capabilities would be able to have a close, personal relationship with you.

It used to be that computing was the most visible industry exhibiting such exponential rates of improvement, but now other segments of the economy are joining this club, including communications, entertainment products, and medicine. Segments of these markets are doubling in price performance every nine to twenty-four months; Internet backbone bandwidth for example, has been doubling every twelve months.[17] In this industry if you double your bandwidth at the same cost only every eighteen months, and not every twelve months, you are destined to be out of business in a few years. We have just begun to experience the impact of these developments. Today most people communicate over the Internet at thousands of bits per second, the same rate used by current telephones. It won't be long, however, before we will communicate at tens of millions of bits per second and beyond, the rate used by high-definition television. Over the next few decades artificial intelligence and other computing and communication technologies will converge in surprising, nonlinear ways to create unexpected opportunities.

Exponential improvement is a property of many knowledge-based activities, like technology, where, as knowledge is accumulated, it leads to increasingly better solutions. For example, the advent of nanotechnology—the ability to design and build devices atom by atom—is beginning to allow materials and mechanical devices to improve at more rapid exponential rates. Biological evolution, which is a knowledge-compounding process based on nanotechnology, has also experienced rapid exponential improvement at times too.[18]

Transition to Knowledge

The exponential economy is driven by the transition to a knowledge-based economy, where one idea builds upon another at increasing speed. Knowledge compounds. Globalization hastens this process by providing more ideas. Ubiquitous, high-speed communication lets us gather those ideas faster. Advances in collaboration software tools, new business models, and improved innovation practices allow even more rapid improvement. It is inevitable that increasingly large numbers of business activities will improve at rapid, exponential rates as they become more knowledge-intensive and are integrated into the global economy. But exponential improvement is hard for us to comprehend.[19] We tend to overestimate exponential impact in the beginning and underestimate it later on.

Consider this folktale. A prince once did a great favor for his king. In return, the king told the prince that he could have any reward he wanted. The prince said he would like an amount of rice computed by a simple formula based on the number of squares on a chessboard. He would take one grain of rice for the first square, two for the second, four for the third, eight for the fourth, until all the squares were filled. The amount of rice would double from square to square. This would continue until all 64 squares on the chessboard were covered with rice. The king thought this was reasonable and granted the prince his request. But when the king learned the result, he was astounded: the amount of rice needed was billions of times more rice than had been produced in the entire history of the kingdom.[20] Indeed, it was nearly enough rice to cover the surface of the earth. Once the king learned he had been tricked, he had the prince's head cut off. There are two morals to this story. First, be careful of things that compound. Second, never embarrass the king.

The exponential developments described above become particularly powerful when they *collide* and create something entirely new, such as the Internet. The Internet is the convergence of computing *and* communications. It is growing and improving exponentially, but even faster than Moore's Law, since it leverages multiple exponential developments. The number of Internet hosts and the content on the Internet are doubling every twelve months. Clearly, the Internet is touching something very fundamental to have created such activity, excitement, and opportunity.

These "colliding exponentials"[21] are creating a major transition in our economy. Fifty years ago most of the economy was improving in price performance but relatively slowly, at a few percentage points per year. Products that are primarily characterized by their physical and material attributes—such as trains, automobiles, houses, stoves, and vacuum cleaners—cannot be improved in price performance as quickly as knowledge-intensive products and services. Other early products, like analog telephones and televisions, could not be improved rapidly because major innovations required completely new standards.

But fifty years ago, Moore's Law for computing was charging ahead. If you were in the computer business during the last century you had to double your performance every eighteen to twenty-four months just to survive. Most did not. Among the first wave of computer companies, only IBM has endured. Electronic Controls Company, Burroughs, DEC, Cray, Honeywell, RCA, Wang, Commodore, and hundreds more are all gone.

In the 1980s, the Internet became technologically and economically practical, and the "law of exponential interconnections" added another exponential driver to the exponential power of Moore's Law. The law of exponential interconnections says that in a fully connected network, like the Internet, the number of connections between users goes up exponentially as users are added, which helps you gather ideas and build on earlier ones faster. This allows exponentials like Moore's Law to go even faster.

Although these early computing and communication developments were remarkable and impacted some businesses, they didn't initially impact the bulk of the economy. Even into the early 1990s there was considerable discussion about whether computer and communication technology had made *any* appreciable impact on office productivity.[22] But as more of the economy became knowledge-intensive, that changed. Today the convergence of both computing and communications has advanced to the point where they are touching most segments of the economy. This includes product sales, auctions, publishing, entertainment, education, distribution of goods, manufacturing, and research.

There is no longer any question about the impact of computers, communications, and other rapidly improving technologies on our economy. But the impact has just begun. The Internet is used by less than 20 percent of the world's population.[23] Within just a few years, more than two billion people from all over the world will use cell phones.[24] And once we have connected the majority of the people around the world, we will then begin to connect our computer devices and applications to automatically share data and knowledge. This is referred to as the "age of ubiquitous computing." It will be possible, for example, for both the hospital and the children of an infirm parent to monitor the parent's health. Eventually, there will be billions of people and hundreds of billions of computers wirelessly connected through networks like the Internet. These expanding capabilities will, inevitably, create wave after wave of surprising new opportunities.

Expanding Impact

As we go forward, exponential improvements will increasingly impact other areas that are important to us, such as entertainment, education, and environmentally green products. Consider, for example, entertainment. Many consumer entertainment products—such as computer games, CDs, cameras, and personal digital assistants (PDAs)—have already become digital. Analog television will be replaced by digital

HDTV in 2009,[25] and soon 35mm movies will evolve into digital cinema. At this point the transition to an all-digital entertainment infrastructure will be complete. The fragmentation and decline of major television news media, national newspapers, and movie studios will continue. The current hierarchical media industry is literally being turned upside down as computer interactive games, blogs, websites, and millions of individuals around the world produce new forms of info-entertainment content. There are, for example, more than 24 million Internet blogs and their numbers are growing exponentially. The increasing emergence of personal communication devices and community-supported information services, such as Wikipedia.org and Craigslist.org, heralds an important new vision of the future.

Medicine and health services will be revolutionized as they fully enter the exponential economy. Medicine has been traditionally a profession based on an understanding of anatomy and basic chemistry. Doctors ask how you feel, take your temperature and blood pressure, obtain X-rays and MRIs, and collect blood and urine samples. From this limited information they try to discover the status of your health. Obviously, compared with those of one hundred years ago, doctors do a superb job. But humans are immensely complicated organisms, and the tools doctors use today are primitive compared with what is needed. We require knowledge of the specific signaling and computational processes occurring within our bodies—knowledge that is rapidly being collected. With the decoding of the human genome, medicine has entered a rich new era based on the detailed processes that determine how our bodies function. Medicine is becoming an information technology. Consequently, we are experiencing a Moore's Law for medicine where, for example, genomic information about humans is doubling every fifteen months.[26]

Personalized medicine will become a reality as we understand how the unique genetic differences in our bodies impact our health. We will take only those drugs that are compatible with our physiologies, eliminating many of the hundred thousand U.S. deaths each year due to adverse side effects.[27] In addition, researchers are now modifying the human genetic code to prevent birth defects and other crippling diseases. Developments in genomics and biocomputation will eventually allow drugs, which now take an average of fifteen years and a billion dollars to develop, to be designed and tested in computers at a fraction of today's cost and time. Life spans will continue to increase, perhaps by ten, twenty, or more years.[28]

As more and more segments of the economy become based on knowledge and leverage the power of the law of exponential interconnections, they too will join the ranks of rapid exponential improvement. Genetically modified foods will continue the revolution in food production. Even industry segments that do not improve as rapidly—such as transportation, energy, and sustainable development—will be positively impacted by these developments. New technologies, such as direct-carbon fuel cells,[29] have the potential to produce more environmentally benign electricity from abundant coal and other resources.

Many traditional products—such as automobiles and toasters—have entered the exponential economy. They have entered not in terms of their base physical characteristics, which improve slowly, but rather in terms of how they are designed, manufactured, and introduced into the marketplace. What determines success is speed of innovation and the new customer value created—the benefits per costs. Increasingly, the competitive advantage in these industries is based on advances in computer software simulation tools, exemplars of the exponential economy. Even crash testing of automobiles is being done through computation, where simulated cars are "crashed" in the computer. In addition, improvements from advances in materials and other technologies are integrated more rapidly into new products as the entire design and manufacturing process becomes based on computer modeling and simulation.

Simultaneously, manufacturing capabilities have continued to improve to the point where the physical quality of many products is no longer the determining factor. Even throwaway consumer products have remarkable quality and features when compared with those of only ten years ago. For example, a plan has just been announced by Quanta to build a robust $100 computer for children, which will have a hand crank to power it.[30] Quality is now required just to enter the game. At the same time, the base cost of many products is now approaching the cost of the materials used. The value of products is shifting from materials and manufacturing to quality, design, new features, and the overall customer experience. Competitive advantage in traditional businesses increasingly comes from exploiting advances in the exponential economy.

Not all jobs are subject to exponential improvement. Barbers cannot increase the speed at which they cut hair by 100 percent every eighteen months. Nor will restaurant staff, hotel employees, and construction workers work at profoundly increased speeds. Some of these workers, those not engaged in knowledge activities, will be less impacted, assuming their tasks cannot be automated or outsourced. But as more and more segments of the economy become knowledge-based, they too will join the ranks of rapid exponential improvement.

Developments within the exponential economy impact segments of business in other ways, even if that activity does not improve rapidly. Google's maps, for example, tell where you can find barbers and restaurants. Automobiles are now very much part of the exponential economy as platforms for information services, such as satellite radio (e.g., Xfm), safety services (e.g., GM's OnStar™), navigation services (e.g., GPS), and a host of other mobile communication and information services. In addition, the exponential economy is forcing major segments of the economy to change their business models and create new customer value. For example, online book sales by Amazon and others are forcing local bookstores to add new services to survive, such as famous author–speaker programs, coffee bars, and comfortable chairs. Wal-Mart is using advances in information technology to tie together a global supply chain that has driven innovation and revolutionized retailing.

And so it goes. Automobiles, retailing, education, travel reservations, media and broadcasting, financial services, and entertainment: all these industries, and more, are trying to figure out how the colliding exponentials all around them will transform their business models and open up new opportunities. Others are working as hard to develop new business models to replace today's incumbents. Just as eBay and Amazon created new business models from the first wave of the exponential economy, the companies that emerge from the next wave of colliding exponentials will both augment and replace the companies that seem so remarkable today.

Productivity

The dramatic improvements in performance and cost we are describing here do not automatically result in correspondingly large improvements in a country's gross domestic product (GDP). The relationship between rapid exponential improvements in knowledge-driven segments of our economy and the overall growth of the economy is complex. Nevertheless, consider productivity, which is the primary driver of GDP. The U.S. productivity increase from 2000 to 2005 was 3.39 percent per year, the highest rate in fifty years. As Arnold Kling notes, "Suppose that productivity growth in the traditional economy is 1 percent per year and that productivity growth in computers is 50 percent per year. In that case, an economy that is 6 percent computers and 94 percent everything else should grow at a rate of 3.94 percent per year."[31] Information technology is now about 10 percent of the total U.S. GDP and increasing at about 0.2 percent per year.[32] We believe he is fundamentally correct and that we are in a long period of increasing prosperity as more and more segments of the economy enter the exponential economy.

How important is this increase in productivity? Nobel Prize winner Robert Fogel has pointed out that the future of Social Security depends on it. He said, "[I]f we continue to grow as we have in the neighborhood of two percent per annum per capita over the past fifty years, we won't have any difficulty paying for it."[33] Three percent productivity growth might even be enough to make Medicare viable. This observation is compelling motivation for making innovation the centerpiece of a nation's economic policy. Government policy will also need to adapt and leverage these megatrends to provide the needed support and infrastructure.

The force of these developments is still in the early days of exponential growth. The ways in which we interact, work, and play will all change, as will our social customs, political processes, and global relationships. Because of these exponential developments, whatever your world is like today, in twenty or thirty years it will be substantially different. We are entering one of the most remarkable, important, and in some ways frightening periods in the history of humankind. The exponential economy has created an abundance of opportunities, but it has also created a profoundly more competitive world. Clearly, as these developments unfold, their impact on the economy and humanity will be important and long-lasting.

Company Lifetimes

One indication of increasing competitiveness in the exponential economy is the declining lifetime of companies, as described in an important book by Richard Foster and Sarah Kaplan.[34] Consider Figure 2.1 from their book, which shows the lifetime of Standard & Poor (S&P) 500 companies over the past seventy years. The S&P 500 is a leading index of large and medium-size U.S. companies. "Lifetime" is measured by how long one of the S&P 500 lasts as an independent company. They can die either by going bankrupt or by being bought and merged into another company. The curve shows two interesting results. First, it indicates regular, and dramatic, business cycles where companies go out of business faster or slower. Second, at the beginning of the twentieth century, companies lasted on the S&P 500 for more than fifty years. Many

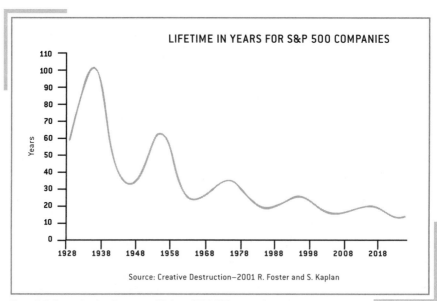

Figure 2.1: Curve showing the average lifetime of S&P 500 companies.

employees had lifetime employment. Today company lifetimes are down to roughly fifteen years. Lifetime employment is effectively gone.

The phenomenon of short, declining lifetimes of companies is particularly astounding because companies that make the S&P 500 are large and successful. They have all the ingredients that are so difficult to put in place: customers, a proven product or service, a business model, technology, capital, people, and brand visibility. But at this point, when everything needed is in place, the average company has only another fifteen years or so to live. Curiously, these companies are not dying because of a lack of opportunities—they are swimming in them.[35] These are just opportunities that the established companies can't respond to fast enough.

These results illustrate that Joseph Schumpeter's forces of "creative destruction" in the marketplace are at work in ways that might surprise even him.[36] The observation about the dramatically shortened lifetime of companies is justification enough to cause us to rethink our innovation processes. As Darwin pointed out, "It is not the strongest of a species that survives, nor the most intelligent, it is the one most adaptable to change."[37]

Failure to adapt can lead to extinction in our world—fast. Foster and Kaplan provide evidence that adaptation to change is extremely difficult for established companies. They point out that over the past seventy years there have only been a few companies able to outperform the market over a sustained period of time. This is significant because if you are falling behind the market, you are probably going away. When Foster and Kaplan wrote their book, they noted only two companies that outperformed the market, GE and Kodak. Since then, Kodak has fallen out of this exclusive club but Johnson & Johnson and Philip Morris may have joined it. Their results indicate that over the long term "built to last" can be achieved but it is mostly an illusion.[38]

A fundamental reason for this failure by established companies to keep up is that they are, by definition, built to fight the last war. Nevertheless, the world unrelentingly changes, demographic shifts occur, customer needs evolve, technologies get invented, management practices improve, and novel business models emerge. The problem is that established companies tend to focus on preserving today's products and business models over emerging, but eventually important, opportunities. They have well-defined organizations and processes designed to achieve these earlier objectives, but these very organizations and processes now resist the changes needed to exploit the new opportunities. Finally, the customer experience and world view of senior managers is based primarily on experiences obtained earlier in their careers, experiences that are now somewhat obsolete. The time needed to overcome these obstacles slows down the speed of innovation within the company. At the same time, the companies in the overall market do not experience these legacy-induced delays. Over decades this inability to innovate at the speed of the market usually turns out to be fatal.

The proof that established companies face a prodigious task in keeping up with the market is that it took Jack Welch at GE, perhaps the best CEO of the past century, to do it. His moniker at the time was "Neutron Jack," which was used by his detractors to denigrate his tough decisions. In retrospect he made the decisions necessary to save the company.[39] He was none too tough.

Andy Grove, the former CEO of computer-chip manufacturer Intel, often said, "Only the paranoid survive."[40] As his comment suggests, there is no room for arrogance or complacency when you are attempting to keep up with exponential growth. That is why we call these developments the exponential economy. We want to stress their rapid and profound effects, which, if ignored, will overwhelm us. The exponential economy means that we live in an extremely competitive world, where one must quickly create and then re-create compelling customer value.

Product Life Cycles

The transition to a knowledge-based, exponential economy has had a profound impact on the product life cycle. Whereas fifty years ago the life cycle of a major new product might have been measured in decades, like the land-line phone in your office, it is now measured in years. For reasons we just mentioned, history shows that established companies find it difficult to make the transition from one major product to the next. The ability to create those new products—the ability to innovate rapidly—is a survival skill in the exponential economy.

In the exponential economy, product life cycles are being dramatically compressed. Consider consumer communication products. After Marconi demonstrated wireless transmission in 1901, the first AM radio broadcast was in 1920 from KDKA in Pittsburgh. They asked their audience, "Will anyone hearing this broadcast please communicate with us, as we are anxious to know how far the broadcast is reaching and how it is being received?" In 1933, FM radio was invented by Edwin Armstrong, and a few years later it was launched. Television was black-and-white from its introduction in 1939 until 1951, when David Sarnoff introduced the next innovation, NTSC color television. The first color broadcast was the Edward R. Murrow program *See It Now,* which showed a picture of the Golden Gate Bridge.[41]

Color television remained essentially the same until 1994, when high-power digital satellite TV was introduced by RCA and Hughes.[42] Once TV was digital, the pace picked up. In 1996, the FCC approved the Grand Alliance HDTV system, which is replacing the U.S. analog system. In 1999, HBO began HDTV satellite broadcast of motion picture films.[43]

Also consider the iPod phenomenon. Just a few years after Apple introduced the audio iPod Shuffle, there are half a dozen competitors on the market, and Apple has now introduced the video iPod. It is just the beginning of a growing family of new media products and services.

Figure 2.2 shows the product life-cycle curve. A new idea is born at **A** and then works its way up in both customer and company value to become a new product or service at **B**. An invention can occur at **A** and be the result of a single individual. But it takes a multidisciplinary team many years to create an important innovation. Thus, the investments required to go from **A** to **B** can be tens to thousands of times greater than those needed to develop the idea at **A**. Unfortunately inventors and the authors who write about them often underappreciate the importance of innovators and the larger team required for achieving marketplace success.

If the new product or service manages to reach **B**, it is introduced into the market and sold to customers. From **B** to **C** the new innovation

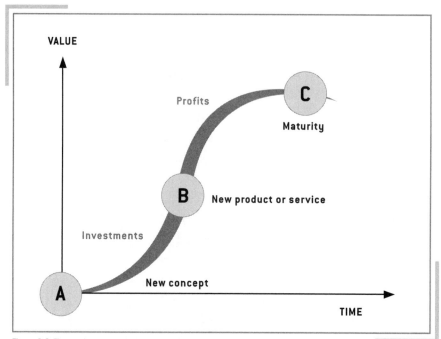

Figure 2.2: The product or service life-cycle curve. It shows a new concept moving up in value to become a product or service and, over time, becoming mature and then obsolete. The process then starts again.

is sold to make a profit. Eventually, the product or service becomes mature at **C** and begins to decline as it is either commoditized or becomes obsolete. The enterprise must then introduce a new product or service to start the cycle all over again.

Philo Farnsworth, the inventor of television, made his primary contributions at **A**. David Sarnoff, the innovator who created television broadcasting, made his from above **A** to **B** and beyond. Few individuals have the skills to make the transition from **A** to **B** and create the teams and organizations necessary for success. Two exceptions were Thomas Edison and Edwin Land. Among other accomplishments, Edison's teams created the entire electrical energy distribution system. Land established Polaroid to make his invention of instant photography an innovation by bringing instant photography to the consumer.

Once an innovation is established in the marketplace, most companies are effective at running their business from **B** to **C**. They have organizational structures in place and well-thought-out processes to monitor and grow the business. The CEO and management team focus on day-to-day operations to keep the company profitable. But a new innovation, which is being moved from **A** to **B,** is often not managed with the same intensity. Indeed, some CEOs consider research an unwelcome cost that lowers the stock price, since it may not pay off on their watch.

Thus, the trouble starts when a product begins to reach the end of its life and a new product or service must be developed and introduced into the marketplace. P&G has a successful process for doing this, and so does W. L. Gore. But this is not the rule. For example, Sony used to be a role model for producing successful innovations, but as their products have matured they have often failed to create the next waves of profitable products. Remember their original Walkman? This huge financial success is now obsolete, replaced by a family of digital devices.

Getting from *A* to *B*

Innovation is the process of getting from **A** to **B** to create and deliver new customer value in the marketplace. This new customer value should also enable a sufficient profit for the company to grow. As we will discuss at length, the conceptual vehicle for getting to **B** is your value proposition, which addresses the following fundamental questions:

- What is the important unmet customer and market need?
- What is your approach to address this need?
- What are the benefits per costs of your approach?
- Why are your benefits per costs superior to those of the competition?

Your value proposition applies at every step from **A** to **B** and is the nucleus of all your activities to create new customer value.

Innovations are not limited to creating new industries, as Sarnoff, Edison, and Land did. The greatest opportunity for people working in established companies lies in making incremental improvements to existing products and services—improvements to products that already exist above point **B.** Incremental innovations have a life-cycle curve that

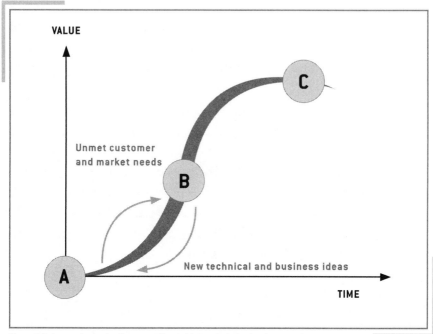

Figure 2.3: Innovations arise from new ideas, either technical or business, and an understanding of important unmet customer and market needs.

is identical to that shown in Figure 2.2 and they need to be developed just like larger innovations. These smaller innovations can be extremely important, especially as one builds upon another. We share the passion that quality gurus W. Edwards Deming and Taiichi Ohno had about the power of accumulated small innovations. A large number of small improvements can create a huge marketplace success, as Toyota and others have demonstrated.

New innovations, whether incremental or transformational, are developed: they don't just happen. Since speed and urgency are essential competitive advantages in the exponential economy, an efficient, disciplined innovation process is needed to quickly seize new opportunities. The "tech push" and "build it and they will come" approaches imply that you simply build a new product and push it up the life-cycle curve from **A** to **B**. This rarely works, because an understanding of what the customer wants is lacking. RCA, which no longer exists, lived and died using this approach. So did Compaq, Polaroid, DEC, and many others. They produced many great technological inventions, but ultimately they did not meet their consumers' needs. Hubris by management coupled with undisciplined innovation practices led to failure and the death of the companies.

The only way to systematically create compelling customer value in the marketplace is to simultaneously interact with both the marketplace and the sources of new ideas, as indicated by the arrows in Figure 2.3. You must continually interact with the marketplace to identify important unmet customer and market needs, and to develop a deep understanding of the overall ecosystem and competition. You must continuously interact with sources of new ideas to understand what is possible, so as to be able to develop new innovation concepts. New ideas can arise from many sources—technology, process improvements, business models, and cost of labor, to name a few. These new ideas are the raw materials out of which an original approach for your new product or service concept is built.

New customer value can be created in many different ways. For example, a new product or service might be based on either a revolutionary technology or a smart product design. But many, if not most, innovations come from the creation of new business models.[44] Consider Southwest Airlines, which might stick in your mind because of their jolly flight attendants. What Southwest actually has is a maniacal focus on lowering costs, not on entertainment. Those jolly flight attendants are to get passengers on and off planes fast to allow rapid plane turnaround. They also use one type of airplane to simplify maintenance, and airports that allow easy access, for the same reason. Speed means one extra flight a day, which results in profitability. Another example is Dell's approach for minimizing inventory by leveraging supplier agreements and by providing customized products on demand.

Principle of Knowledge Compounding

Why do so many knowledge-based activities improve exponentially? Although the reasons are generally complex, a simple explanation based on the compounding of knowledge can give important insights into the process. It will help you understand when you should expect exponential-like performance improvement and when you should not. This knowledge-compounding model also plays a central role in our value-creation processes for innovation teams. If you can't innovate at rapid, exponential rates, you cannot thrive in the exponential economy.

Einstein is purported to have said, "The most powerful force in the universe is compound interest." If you understand this principle, you can begin to grasp the exponential economy and why it is so important. Consider a simple example we are all familiar with that illustrates the basic concepts: If you have a dollar today and the interest rate is 15 percent per year, your dollar will be worth $1.15 in a year. The following year you will have $1.15 × 1.15 = $1.32. After five years your dollar is worth $2.01—you have doubled your money. In a hundred years, by starting with only one dollar, you would have more than a million. And if your bank compounded your starting dollar at 50 percent *every day,* you would be a millionaire in 35 days. So the proper conclusion to

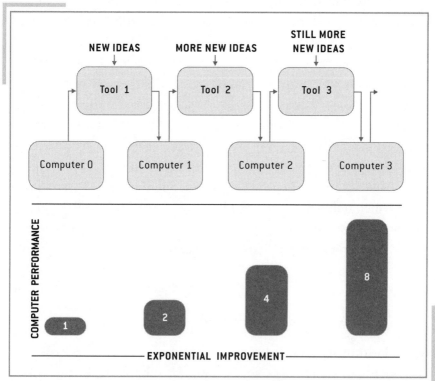

Figure 2.4: Top: Tools, new generations of computers, and the compounding effect of the accumulation of new ideas. Bottom: What exponential improvement looks like. In this case, the performance increases by two times per cycle.

draw from this analysis is to make the interest rate large and to compound often. That's what loan sharks do.

We once ran out of cash on a vacation trip after the banks closed for a holiday. We asked the son of a friend, five-year-old Peter Winarsky, if we could borrow the $20 he had brought on the trip. It was his entire life's savings. Peter looked at us in disbelief. So we made a deal with him. We said, "Peter, we will give you ten percent compound interest for *each day* we keep your money." He asked, "What is compound interest?" After it was explained to him, he looked off into space for a long time while he did some mental math, and then turned to us and said quietly, "I like this idea. You can borrow my money." Smart fellow.

Technological and business developments can lead to exponential improvements too. Each new discovery builds on previous discoveries to create additional improvements. As long as there is an important *need* for the resulting innovation, *new ideas* to drive the innovation process, a *compounding* process to let improvements build on previous results, and the financial, human, and other *resources* to make it happen, exponential improvement can occur.

Figure 2.4 illustrates this process. Imagine, as an example, that a computer is used as a tool, Tool 1, to collect new ideas to design a new computer, 1. That computer then becomes a new tool, Tool 2, to collect still more ideas to design another new computer, 2. This process can repeat for as long as there remains an important unmet need for improved computers, new ideas, a way to keep building on the earlier developments, and appropriate human and other resources.

In simple terms, this is the engine that keeps Moore's Law going or, for that matter, any knowledge-compounding process where these basic conditions are satisfied. The formula for a compounding improvement process is identical to the formula for compound interest.[45] As with compound interest, you want to collect as many great ideas as you can and compound often.[46] There is evidence that Moore's Law, for example, is speeding up and will soon exhibit 100 percent improvements in price performance in less than twelve months.[47] It is likely that many other activities will speed up too, as more people around the world add new ideas and as an improved communication infrastructure allows faster sharing and compounding of ideas.

In *Through the Looking-Glass,* the Red Queen said to Alice, "It takes all the running you can do to keep in the same place. If you want to get somewhere else, you must run at least twice as fast as that!"[48] As Grove's earlier comment implies, being in the computer-chip business, or any other rapidly compounding knowledge business, is a little like being in Wonderland. The Five Disciplines of Innovation are about how to "run even faster."

Application to Innovation Teams

The ideas discussed above apply to any improvement process, and if you are working in a profession that is improving at exponential rates, you need an *exponential improvement process* to keep up. That is, if you are involved in an activity where you are accumulating knowledge to make something better, you want to

- Work on an *important* need, where you can stay ahead of the exponential wave of improvement in your activity.

- Have access to *many new, valuable ideas* to make the largest possible improvement at each step.

- Employ an *iterative compounding process* so the magic of compounding can occur.

- Have the appropriate financial, human, and other *resources to drive the process.*

The goal is to achieve exponential improvement that will put you ahead of the exponential improvements occurring in your activity by your competitors. These ideas are at the core of our value-creation process: important customer and market needs; tapping the genius of the team for ideas; tools and processes to compound these ideas to build new solutions quickly; and champions and innovation teams to drive the process. The more successfully we emulate these conditions, the more rapidly and successfully we can innovate to create new customer value.

These concepts can be applied to many different activities, including product development, quality control, government policy, education, and basic research. They extend and enhance traditional "brainstorming," where groups get together and develop new ideas. They extend and enhance the pioneering contributions on quality improvement by W. Edwards Deming and Taiichi Ohno, but now applied to value creation for knowledge workers. Because the ideas described here are general concepts, they are effective even if the activity you are working on is not changing as rapidly as in, say, the computer business. This includes all creative and innovative activities.

Although the exponential economy presents many challenges, it also represents a time of unprecedented opportunity. The emerging global marketplace, plus wave after wave of exponentially improving technologies, will continuously create major business opportunities in most market segments. We do not agree with the "doom and gloom" crowd who only see problems ahead. Rather, we believe that these global developments will produce economic improvements around the world, including the United States. In 2005, for example, worldwide poverty rates significantly declined. In China, poverty is declining by one million people a month.[49] These positive developments should continue and even accelerate for the foreseeable future.

As Einstein suggested, events that compound can create surprisingly impressive results. We are solidly in the exponential economy. We need to heed Einstein's observation and master the most powerful force in the Universe.

DISCIPLINE

1

important needs

CHAPTER 3

WORK ON IMPORTANT CUSTOMER AND MARKET NEEDS:
the RfID tag

"WORK ON WHAT'S IMPORTANT, NOT JUST WHAT'S
INTERESTING—THERE'S AN INFINITE SUPPLY OF BOTH."

Frank Guarnieri

Precious Metals

During one of SRI's regular innovation meetings, Sunity Sharma and his team gave a marvelous presentation as part of an effort dedicated to exploring new printing technologies.[1] Amazingly, they even demonstrated how to print metals such as gold, silver, and platinum on normal bond paper. At first blush, this seems impossible. How do you melt gold and print it without the paper burning up? Sharma explained that his team had invented an alternate, extremely inexpensive process costing only pennies to coat a page of paper without heat. His sample page and prototypes sparkled like a mirror, building excitement that this technique could make the most striking, distinctive greeting cards imaginable.

After the initial euphoria about what seemed like a real breakthrough, we asked a hard question: "Does anyone care?" Unfortunately,

Sharma's team, through market research, found there were already good alternatives for making shiny surfaces on paper and that, even if there hadn't been, the market was only a few million dollars per year. While the new technology was amazing, metallic greeting cards remained an interesting business opportunity rather than an important customer and market need.

Later, a coincidence happened that connected with the printing technique. One of our business development staff was investigating the product logistics market. Every product that is produced needs to be shipped somewhere. Estimates indicate that while only a small percentage of shipped packages are misplaced at some point in their journey, the total cost of keeping track of packages was worth many tens of billions of dollars per year[2]—an important market.

One way to track packages is with an emerging technology called RFID (*R*adio *F*requency *Id*entification) tags. An RFID tag can be about the size and shape of a small Band-Aid. When activated by a radio frequency pulse, it emits a radio frequency signal containing information about what is in the package and where it needs to go. The signal can be read with a small monitoring device at a distance. No one needs to touch the package. Imagine, for example, thousands of packages coming off a ship from Japan on a conveyor belt, each package with an RFID tag. A machine would automatically read the tag and then send the package off to its appropriate destination. Because of the RFID tag, you could locate the package at all times. The market for RFID tags is estimated to be almost a hundred billion units per year by 2010.[3]

So what's the problem? For the tags to be put on enough packages to justify installing the system, they must be very inexpensive—less than 5 cents each. If you examine what's in an RFID tag, you will find that there are only two major parts: a small processing chip to hold the information about the package and its destination and an antenna to receive and send that information. The antenna is only a few inches long but it costs about a dime today because it must be made out of high-quality copper. What is needed is a way to print high-quality copper antennae on paper at an extraordinarily low cost. If antennas could be produced for a fraction of a penny, with the production of hundreds of billions of tags per year, it would create a potential market of several hundred million dollars each year.

This is an *important* market need Sharma's team can fill, not just an interesting one. In five years or so, we will learn whether their technology is the winner in this marketplace. Unlike the greeting cards idea, this opportunity is worth the effort.

This example illustrates the first of the Five Disciplines of Innovation: focus your efforts on important customer and market needs—not just interesting ones.

When you focus on important needs, it is possible to obtain the financial resources necessary to achieve success in the rapidly improving exponential economy. For instance, no one was going to finance a technology to make more attractive greeting cards, but a revolutionary technology that addresses a major need in the huge RFID market attracts investors like bees to a honey pot. Important needs also allow you to attract the best people, who are excited by the opportunity to make a contribution. Important customer and market needs are also critical for making innovations that will endure long enough in the marketplace to allow for financial success.

Our complex, global world offers a huge number of fascinating, unsolved problems. Selecting an important unmet customer and market need at the right time is the critical starting point for all success. Then you must make sure your concept is feasible and the necessary infrastructure is in place to make your project commercially viable. Even then, if you can't raise the resources needed, the journey is futile. By answering the four questions in your value proposition, as we will do in Chapter 5, you will be able to tackle these issues and create timely innovations.

Sharma and his team were able to redirect their focus to an important problem, but what happens when people focus on interesting, unimportant problems? The faculty at colleges and universities, for example, rarely discuss how to encourage their members to work on important problems. Rather, each individual faculty member makes a decision about what to work on and forges ahead. As a result, university faculties are often accused of being irrelevant to the needs of society. Rather than having six or sixty faculty all pursuing projects unrelated to one another, it is possible to have faculty sharing ideas on a regular basis to build higher-impact programs.

While multi-disciplinary collaboration is the exception at most universities, major universities like Stanford are beginning to succeed at it.[4] But collaboration in universities is generally hard. I was, for example, having an extended conversation with the dean of engineering at a top-twenty U.S. school that had just created a new center for joint research in nanotechnology.* I asked him how it was going, and much to my surprise he blurted out, "We are going out of business. We can't get people to work together in ways that are required today to solve hard problems. As soon as we formed the new center with the objective of pulling together the needed disciplines, everyone immediately went back to the old stovepiped way of doing business." Obviously, this great university is not going out of business, but the dean's comment represented his frustration at trying to get the faculty to collaborate.

As I listened to the dean tell his story, I was thinking that he had little chance of success. He was attempting to apply one of the Five Disciplines of Innovation—work on important problems—without taking all five into account. He was failing, in part, because the university was not aligned with his goal. The reward systems in his university, such as getting tenure and salary increases, recognized individual contributions, not team performance. But in this situation extraordinary performance could only come with collaboration in multidisciplinary teams.

As described in Chapter 2, we are interested in improving the success rate of innovations over the entire continuum, from incremental to transformational. When RFID tags are eventually applied across the logistics industry, it will be a transformational innovation that will rearrange significant parts of the logistics industry. Developing a low-cost RFID antenna, as Sharma is doing, is an important innovation because it is a critical part of the overall RFID system. Incremental innovations—such as a new financial tracking system in a company or a new product design— are much smaller, but the principles are always the same: Focus on what is important to your customers and not just what is interesting to you.

---

*Referring to coauthor Curtis R. Carlson.

Criteria for Important Customer and Market Needs

In all enterprises the goal is to use scarce resources, whether financial or staff, to create the greatest customer and shareholder value. But in any organization it is almost impossible not to be distracted by all the "interesting" things that can be done. Every organization must quickly sort through the hundreds of possible projects and select those that will make the greatest impact. This selection process, in conjunction with the other disciplines of *Innovation,* is improved by having commonly understood criteria across the enterprise for what constitutes "important" customer and market needs.

It is the responsibility of the senior management to develop these overall criteria. Then employees at every level should understand these criteria and be able to identify important needs within their activities. Unfortunately, most organizations have neither the criteria nor processes to establish what "important" means to them. The result is waste and failure.

We have found that just having the *expectation* of doing important work in an enterprise has enormous benefits because it focuses the enterprise's energy and thus improves productivity. Consider the Frank story from Chapter 1. When Frank described his technology for developing new drugs faster, he immediately satisfied the important-need criterion. As a consequence, it was an easy decision to pay attention to him and figure out the merits of his proposed innovation. And, of course, important projects are almost always interesting too. Frank's certainly was.

It is fascinating to go into an organization and ask managers to describe their projects and tell you, quantitatively, what the impact of success would be for the enterprise. Rarely will they be able to quantify the need, the resulting customer value, the status of the competition, and, assuming success, the overall impact. In many cases you discover that success would not make any significant difference to either the customer or the company. A project that seems interesting to do often turns out to be unimportant—a waste of time and resources. Neither the employee nor the enterprise have systematically evaluated their projects.

In the SRI Discipline of Innovation workshop, we discuss the principle of addressing important customer and market needs and not interesting ones. During one workshop a manager from a major broadcaster stood up and announced to her colleagues, "I now realize that my project, while interesting, is not important in providing value to the corporation. When I go back home I am going to suggest that my project be disbanded and that we focus on the important new market arena of content delivery to handheld devices." We all admired her courage. Such shifts, centered on the importance of the project to customers, are hard. But her decision will allow her to spend her energy on an idea that will reap rewards for her customers, the enterprise, and her.

Government agencies are notoriously poor at focusing on customer needs that matter. We are somewhat sympathetic; they are often in a no-win situation since they must deal with hundreds of different constituencies with conflicting needs and objectives. Yet the Federal Emergency Management Agency's (FEMA) slow response to Hurricane Katrina in New Orleans is an example of not having a clear focus on important customer needs. What could be more important than an action plan to evacuate and care for hundreds of thousands of people who are living below sea level in a hurricane alley protected only by a dirt levee? But, amazingly, on the trip to New Orleans rescue workers had to stop for "sensitivity" training,[5] which was certainly an interesting thing to do when compared with the profoundly more important task of saving lives.

There is, however, one U.S. government agency that has clear criteria for what constitutes important needs: the Defense Advanced Research Projects Agency (DARPA). As part of the Department of Defense, DARPA funds several billion dollars of advanced research each year. DARPA is very clear about its objectives. All programs must provide a significant impact on the performance of the Department of Defense by creating paradigm-shifting technical capabilities, which often means an improvement of ten times. For example, DARPA funded the work that linked computers in a network, which led to the Internet.[6] And when you go on a hike and tell your spouse your GPS location in case of emergency, you are using DARPA developed technology. DARPA's mission is to "enable radical innovation in support of National Security."[7] Note the word *innovation* and not *creativity* or *peer-reviewed papers*. DARPA is successful only when the agency's work is used by its customers. This bold mission has resulted in the most productive U.S. government R&D agency, which has propelled such military innovations as stealth aircraft, unmanned combat air systems, autonomous robots, and the Saturn rocket.

Important External Customer and Market Needs

Depending on the purpose of a particular team and enterprise, new innovations can span the space from incremental to transformational. In addition, the criteria for success will vary greatly depending on the nature of the enterprise's business.

If you are a venture capitalist in Silicon Valley, for example, the criteria are well understood. You will be interested in business opportunities with revenues in excess of several hundred million dollars per year and with ROI's of more than 15 percent per year. Smaller revenues and returns on investment will make it impossible to raise the capital needed, hire the quality of staff required, and make the financial returns investors expect.

Major new innovations in established companies must satisfy similar strict criteria. To be part of GE, for example, Jack Welch made it crystal clear that a division had to be in a rapidly growing, multibillion-dollar business and be number one or two in its market segment. Smaller companies will have different criteria but they should be similarly spelled out.

At SRI we have criteria for our spin-out companies that are the same as venture capitalists. But we have very different criteria for new research areas. Here we start by looking at critical societal needs, such as clean energy, environmental remediation, K–12 education, next generation wireless systems, cyber security, cancer, infectious disease, addiction, and aging. Within these areas we cultivate opportunities where we can build Centers of Excellence with research teams of more than 50 staff members.

Important unmet customer and market needs are generally characterized by markets with large revenues. New products and services, such as online retailing, low-cost titanium, RFID tags, and high-performance solar cells satisfy this criterion. Important customer and market opportunities can also impact a large number of customers, exemplified by satellite broadcasting of digital radio, cell phones, entertainment products, and online education. When a drug company produces a new therapy, like Lipitor to lower cholesterol, the potential number of potential customers and the sales revenues are both huge. Of course the investment required is huge too, perhaps over a billion dollars.

Consider another specific example, corporate research and development (R&D), where the fundamental criteria for evaluating programs are customer value, market size, investment, time to market, and return on investment (ROI). These criteria should be quantified and widely understood. But few R&D organizations have disciplined processes based on these criteria. One exception is GE, which is uncompromising about addressing important needs and demanding a solid return on its R&D investments.

Most people working in established organizations are not developing "the next big thing." Rather they are working hard to make improvements to existing products and services, whether through cost reductions or new services and features. Their achievements may be incremental innovations, but they are fundamental to the enterprise's success. The Schick Quattro, for example, was an important incremental innovation to keep up with the Gillette Mach 3 razor. In every job there are a host of things you can work on, but the objective, again, is to address your customer's most important needs—not just the ones that are interesting to you.

Important Internal Customer Needs

Everyone in the enterprise must be focused on creating new customer value. This includes core functions, such as human resources, computer services, and finance. If you are in one of these departments you are not selling products directly to the marketplace. Rather, your activities are meant to provide the best value to your internal customers so they can then provide the maximum value to the enterprise's external customers.

The criteria for core enterprise functions are determined with senior management and the internal customers. For example, staff recruiting and retention is a critical HR function. At SRI one metric is to have a *voluntary* staff turnover rate that is half that of the company average in Silicon Valley.

The failure to address the operating staff's most important needs is a central reason why there is often smoldering antipathy between the operating staff and the corporate staff in some companies. For example, if you are the manager of a company's computer services department and you roll out new management software, unless it helps the enterprise better address important customer needs, it is not a success. Core internal departments are often told to cut costs to improve productivity. As we have all experienced, that can be done at the expense of the staff. The goal is actually to cut costs *and* increase the productivity of the staff. The word *and* is the key to success.

The objective is always to think about how each function of the enterprise can deliver additional customer value. Consider, as an example, Patrick, a researcher in a large organization. He shared with us two experiences about an important problem he had—trying to rapidly hire a new person so his team could complete a customer project within a short time line. Patrick calls to the HR person in charge of hiring:

PATRICK: Hi, this is Patrick from the education group. Our big client wants to augment our current program, but to complete the additional work we need to hire someone right away.

HR: Have you filled out Form 1023?

PATRICK: I don't know Form 1023, but we need to move fast on this. We have identified the person, and the client wants the work to start immediately.

HR: Are you saying that you refuse to fill out Form 1023?

PATRICK: Not at all. We just need to move fast.

HR: Have you advertised the job nationally?

PATRICK: Well, no, but we have someone and the client agrees. So we have to move fast.

HR: You may need to move fast but that doesn't mean that I should not do my job.

PATRICK: No, no—whatever. I'll fill out the form. How soon can you get it to me?

HR: Do you know how many people work here? Do you think I can just drop everything and take care of your problem? Hundreds of people are asking me to do things.

PATRICK: Okay, I'll come get the form.

HR: I'm going to lunch now. I'll put it in interoffice mail. What is your room number?

PATRICK: No, no. Let me come and get it.

HR: You just don't want to follow any rules, do you?

Fast forward to the new HR person:

PATRICK: Hi, this is Patrick from the education group. Our big client wants to augment our current program, but to complete the additional work we need to hire someone right away.

HR: Great, which project?

PATRICK: The California one we have been working on for the past few years.

HR: Great—that project is really turning out to be a good one. Okay, do you have someone in mind already?

PATRICK: Yep, someone the client knows. She is perfect for the job, and she can help on some other projects.

HR: Okay. Give me the details now: level, salary range, and basic job description. I'll write up the job announcement and bring it over right now. You sign it, I'll take it to your supervisor, and we'll have it posted today.

PATRICK: Wow, what should I do?

HR: I'll take care of the paperwork but you need to check references, document those, and make sure this is the person you want.

PATRICK: Okay, but time is of the essence.

HR: We have to post the job: that will happen today. But we can move forward while it is posted and get this person on board ASAP. Have the applicant get in touch with me and I will make sure all the paperwork is in order and make sure she gets all the information she needs to make her decision. There's no reason we can't make this happen in a few days or so.

PATRICK: Thanks!

It's obvious which HR person responded to Patrick's concern, thereby creating both customer and enterprise value. This is a simple example, but many people working internally in organizations are not thoughtful about ways that they can provide maximum customer value. When they are not focused on important needs *for their customers,* they underachieve. Not only are these folks a potential nuisance, like the first HR person Patrick consulted, they diminish their value to the organization and its customers.

Other Considerations for Important Customer and Market Needs

When developing a new market-based opportunity, "importance" includes two other criteria: it should not be overwhelmed by other exponential developments, and it should be feasible.

1. *The opportunity will not be overwhelmed by other developments.* An innovation that does not anticipate external changes may fail. In the

exponential economy, progress is so fast that we must avoid projects that will rapidly become obsolete as they are eclipsed by the "new, new thing." The computer industry, for example, has a huge burial ground for specialized computer chips that tried to surpass the performance of the general-purpose chips that go into personal computers. But the 100 percent cost-performance improvement every eighteen months that exists for general-purpose chips represents an incredible rate of progress. Getting out in front of that rushing train is a formidable challenge. Most get run over.

The exponential developments around you can also result in challenges to you from competitors using better business models. For example, if you are in the paper book distribution business, there is an immediate threat from digitized books and databases, which will make many hardbound books less compelling to customers. For example, Amazon.com is experimenting with the latest generation of e-books so that they can directly download material—"iPods" for readers. In addition, the University of Phoenix is working to do away with conventional textbooks and replace them with e-books. These innovations provide lower costs and greater convenience for their customers. Clearly they are just the tip of the iceberg. Consider that encyclopedias, which cost $1,000 twenty years ago, are now available free over the Internet. Rapid obsolescence is the rule.

As another example, the electronics and retail industries are facing new competition as companies emerge from China and as new, worldwide production and distribution systems are perfected, such as those exemplified by Dell. Wal-Mart has further leveraged these developments to create a transformational retailing innovation.

Regardless of the rate of change for your own products or services, it is as a colleague says, "Aim ahead of the curve." Failure is a waste of time and resources.

2. *The opportunity is feasible.* Innovations are possible only when the product or service, infrastructure, and other inventions and resources required can be put in place. At any given time, there are a million ideas that can never lead anywhere. Leonardo da Vinci, for example, was a prolific visionary who described the helicopter, hang glider, ma-

chine gun, tank, submarine, mechanical calculator, and solar heater. But in these cases he was neither an inventor nor an innovator, since his exciting ideas were impossible to build at the time. His ideas provided no additional customer value.

Sometimes an innovation is impossible because the technology is still too expensive (e.g., RFID tags and fuel-cell cars). Sometimes, it is impossible because an appropriate business model is missing (e.g., digital cinema). In some cases it is because the people with the idea don't understand how to proceed. About fifteen years ago, a friend, who spends many snowy days out of doors, identified the unmet need for portable hand warmers based on chemical reactions, but he didn't know any business people capable of manufacturing and distributing such a product. As a result, this potential innovation didn't lead to anything. Others who knew how to create the product and distribute it now have their products in stores.

Thomas Edison was a master at knowing when to attack a problem, and he created one marvelous innovation after another. For many years he did not work on the lightbulb. He realized it was a huge opportunity, but he also knew that the technology needed to create the infrastructure to distribute electricity was not practical. Once he decided the infrastructure could be built, he put his prodigious energy behind the task and created a durable lightbulb, parallel circuits, an improved dynamo, an underground conductor network, safety fuses and insulation, and light sockets with on-off switches.[8]

The objective is, as Edison knew, finding solutions to important customer and market needs where all the pieces can come together. The electric lightbulb was an invention. The creation of the electric lightbulb with a practical electrical distribution system that could economically deliver power to customers was an innovation—one of the most important in the history of mankind. Edison did his homework and, among many other things, created the phonograph, the cinema, and the modern research laboratory. He also created General Electric to build his innovations, now the second most valuable company in the world after Exxon with a market capitalization of roughly $300 billion.

Consider a contemporary example. Before the human genome was decoded, the ability to create patient-specific drugs for addiction was limited. Now that the basic research on decoding the human genome has been completed, it is becoming possible to understand a person's individual genetic predisposition to addiction, which may eventually lead to specific cures for each individual.

In order to know whether you are focusing on an important market innovation, you need to do your homework. Clearly identifying those unmet needs at a point when a solution is possible is critical for success. Customers must be ready for your new concept, and all the elements necessary for a solution must be within reach.

Selecting Your Important Projects

Look at your own projects and apply the first discipline with these tests of significance in mind:

1. *Do you work on innovative activities that can make a demonstrable difference to your customers, whether internal or external?* Can you answer the "so what" question? Will your activity positively impact external or internal customers and cycle back and help the organization itself? If you feel your ideas are not valued in the organization, maybe the activity has to be refocused so that your ideas *do* matter.

   To clarify the criteria for "important" needs in your activity, write down an initial list and take it to your supervisor. If there is confusion or disagreement, begin the dialogue to reach agreement.

2. *Is it aligned with your organization's goals?* If you are working at cross purposes to your organization's needs, why should it support you? You need to either find a way to make it align or drop it. If you work in a large oil company that is only interested in billion-dollar initiatives and your ideas are for new clean energy systems that are worth millions, you probably need to go elsewhere. If you work in a non-profit organization and continually propose large, unaffordable projects, they cannot result in completed innovations.

3. *Are you willing to commit to it?* Remember again Frank in Chapter 1? He was a perfect example of how commitment is required. If you are not committed to your project or invention, it will never translate into an innovation. It is easy to have a momentary stroke of genius; it is much harder to dedicate yourself to bringing an innovation to fruition through years of hard work.

How about you? Are you working on a project that's worth your time, your company's resources, and your customers' money? Do you have criteria that establish what "important customer and market needs" mean in your enterprise?

The starting point for systematic, successful innovation is the identification of important market and customer needs, whether in the research department, marketing and sales, human resources, or finance. In the exponential economy, all of us must spend our time making the maximum contribution to creating customer value. It's a matter of survival. Whether a large or small business, university, nonprofit organization, or governmental agency, when the entire organization is focused on solving the most important customer needs, an exciting, empowering workplace is created.

To repeat what Frank, who created Locus Discovery to rapidly develop drugs, pointed out, "Work on what's important, not just what's interesting—there's an infinite supply of both."

CHAPTER 4

CReatInG CUstomeR vaLue: *YOUR ONLY JOB*

"I said to my dad, 'I am just one good idea away from
starting my own venture.' My dad said, 'No, what you
need is not an idea, it's a customer.' "[1]
Sarah Nowlin

"Customers are sacred."
Norman Winarsky

Customers?

"Did we hear you correctly? You lost ten billion dollars?" Norman
Winarsky, the vice president of ventures at SRI, asked Bob, a managing
partner from a highly publicized Internet start-up incubator company.
We had begun the conversation by asking how his company was doing
and, even though the Internet bubble was coming to an end at the
time, his answer took our breath away.

"Yes, over the past eighteen months we lost nine billion dollars in equity and a billion dollars in cash." When we asked what had happened,
Bob said they began to believe they were invincible and that they knew
better than their customers what was going to work. Bob said, "We just
kept on investing more money without doing the hard, essential due
diligence. It seemed obvious to us that all our new dot-com companies
were going to continue to go up forever. It was a classic case of hubris."
After more discussion Norman pointed out that their decisions were

being made on the basis of how many PCs they could get access to rather than whether those PC owners wanted to buy anything. Not enough did. Bob is a smart, savvy entrepreneur, but the excitement of the bubble distracted him from the essential hard work of understanding and addressing important customer and market needs—not just interesting ones. While the company got in trouble and had to be restructured, Bob bounced back. Smart people usually do, especially when they learn important lessons like understanding customer value. Today he has a senior job with a prestigious communication company in Silicon Valley.

The hard lesson that, unfortunately, must be learned over and over again is that you don't define value—your customers do. Whether it is the Ford Edsel, the Apple Newton, the Xerox Alto, or the thousands of other products and services that failed and that few can now remember, you always need a customer. The bubble was a fever swamp where the customer was too often forgotten in the rush to get rich quick. The list of companies that initially made a splash but disappeared is legion. Webvan, Pets.com, and eToys are just three examples of hundreds of companies that failed.

Everyone has customers. Whether you are an actor on a television show, a member of a government agency, a salesman, a researcher, an educator, a writer, or a priest, customers are critical because they define your success when they buy and use your products and services. They demonstrate with their actions whether you are creating value or not.

Many people misperceive what customers want, never asking whether their needs are being met. For example, if you manage the local Meals on Wheels program, you will want to know how your clients respond to the timeliness of delivery, taste, and other aspects of the food important to them. You can't guess what customers want, nor can you just use your own preferences as the guide. You may be surprised to find that the most important benefit for some Meals on Wheels clients is having a short, sympathetic conversation with the delivery person.

Unfortunately, some people in companies can't even tell you who their customers are. Identifying your customers and working to uncover their needs is the first, most important step on the path toward maximizing customer value.

Value Creation Is Everyone's Job

Since *everyone* in an organization has customers, everyone should have a clear understanding of the different dimensions of customer value and how their efforts contribute to the organization's success. Ask people in big companies, government organizations, and universities for the value they are providing and you seldom get a straight answer. For example, who are the customers of a tenured professor at a university: the students, the students' parents, the wider public, the administration, the Board of Regents, or the professor's colleagues?

In a remarkably candid comment in a *New York Times* story, Ellen Willis, professor and president of the American Association of University Professors chapter at New York University, was disappointed that increasingly "students and their parents are seen as customers to be satisfied."[2] She implied that, somehow, the teachers are the ones who should be considered the customers! Excuse us, but who is paying the bill? Willis's attitude makes us wonder what education will be like when organizations around the world learn how to compete for tens of millions of students each year over the Internet and in new, for-profit schools. Since the education market is over $200 billion a year in the United States alone, someone is going to figure out how to treat students as the customers they are, provide superior value, and change the way we learn. This shift is already beginning, as illustrated by for-profit companies like the University of Phoenix and the increasing availability of online courses.

The University of Phoenix, accredited in 1979, is the largest private university system in the United States, with more than 160,000 students attending classes at over forty campuses and with a growth rate of 30 percent per year.[3] Phoenix's customers are primarily adult learners. Founded by a former history professor, John Sperling, Phoenix has shaken up the educational community. Sperling eliminated tenure and all lectures. He developed a standardized course curriculum built around peer-based learning in groups. Sperling says, "The faculty member is an equal in the classroom. His job isn't to expound wisdom; it's to serve a learning group." Sperling is now working to do away with textbooks, using electronic notebooks instead. Phoenix has authors create course materials to its specifications, which allows it to bypass traditional textbook publishers and eliminate the cost and logistic hassles of distributing more than a hundred thousand books every few months. Combine these innovations with a growing, worldwide online capability and you sense a sea change in adult education.

As we noted with Bob at the start of this chapter, the dot-com bubble of 1998–2001 was the result of companies being formed without understanding their customers' needs. We would ask people in these companies to describe the value proposition for their customers and, in turn, how that would allow them to make a profit. They would say something like, "We'll figure that out later; our focus is on getting eyeballs looking at our website." Eyeballs? Eyeballs! By 2001, hundreds of companies, thousands of employees, and tens of billions of dollars were gone. It's a rule: you should be able to easily understand the value proposition for any company that aspires to be successful. If you can't, stay away.

Many jobs in an enterprise do not deal directly with external, paying customers. These employees have "internal customers." For example, a lawyer working for the managing director of a product division might have the task of drafting contracts for outside vendors. Does the managing director feel like a customer? Has he ever been asked to prioritize his needs? He can certainly say what his three or four most important needs are and whether they are being satisfied. Few companies measure internal customer satisfaction of people like the managing director.

We once asked the chief scientist of a high-tech start-up company, "What's the customer value that will result from the product you are building?" He mentioned several product *features,* such as "We have really fast processing speed and a great user interface." When we asked whether these features were ones his customers cared about the most, he replied that he thought they were, but he really didn't know. Neither did he know whether these were the most important features compared with those of the competition. At this point we thought, "How can he lead his team and make the dozens of decisions necessary every day, when he doesn't understand his customers and the value proposition for his product?" Soon after our conversation, his company went out of business.

Another time, while working with a major international organization, we asked a group of mid-level managers for their project's value proposition. To our amazement they said, "We work hard." We agreed that they worked hard and explained that we were interested in knowing about the customer value they were creating through their hard work. Again we were told, "We work hard." These were smart people. The problem was that their organization was not based on an understanding of customer value. We were literally speaking a foreign language to them.

The objective of all healthy enterprises is to constantly strive to create greater customer value, first by understanding the customer's most important needs and then by addressing these needs by rapidly creating compelling customer benefits at lower costs. This is your only job.

Types of Value

There are many meanings of *value,* so there is often confusion about how to think about it and deliver it to your customers. All organizations must address at least:

- customer value
- company value
- shareholder value
- employee value
- public value

Customer value—that is, *value to the customer*—is the center around which all other types of value revolve. You can't develop the other types of value until you understand your customers' needs and decide whether or not you can create a compelling product or service for them. If a CEO talks exclusively about increasing shareholder value, he or she provides little or no guidance to employees about how to do their jobs better. All employees, especially the CEO, should spend the bulk of their time talking about their customers and how to address their needs.

However, if you produce products with high customer value, but they cause the company to lose money, the company won't be around long. Consequently, if you are launching a new product, you should construct at least two value propositions—one focused on your customers and one focused on your enterprise. The next chapter provides a template for these value propositions.

Shareholder value flows from delivering high customer value in a large, growing, profitable market. In a publicly traded corporation, shareholder value is a primary concern of the CEO, board of directors, and upper management. It is factored into all strategic investment decisions. Yet a company that emphasizes shareholder value at the expense of customer value eventually fails, as DEC did.

Employee value involves salaries, bonuses, and other reward programs. But it also includes the nature of the work, the quality of colleagues, and what the organization values, plus the enterprise's location, facilities, and other amenities. It includes the enterprise's stability and future prospects, in addition to opportunities for professional growth and achievement.

Public value recognizes that enterprises have both legal and moral obligations to support their communities. A company could, for example, maximize its return at the expense of the environment, which would negatively impact overall public value. The government sets the minimum rules in these situations. These rules come in many forms— taxes, work regulations, environmental standards, and others. An innovation must not only create customer, shareholder, and employee value, it must satisfy the needs of public value as well. Most companies, like HP, take a more comprehensive view of their obligations to their communities, going beyond the minimum legal requirements. Good corporate citizenship includes supporting local education, transportation, and community activities, such as sponsoring the neighborhood soccer team, just to name a few.

State-owned corporations, like the British Broadcasting Corporation (BBC), confront the issue of public value directly. Working with us, the BBC decided that its objective was to create the maximum "public value." As you might guess, this is not easy to do given the BBC's varied constituents, since they vary from passive TV viewers, to bloggers, to cutting-edge, independent content providers. The BBC's new mission—to maximize public value—has the potential to revolutionize the way we measure the performance of public enterprises by fundamentally altering all aspects of their operation, from how they decide on new initiatives to the treatment of customers.

Because there are so many types of value, it is important to clarify which type of value you are discussing. Throughout this book we focus on customer value because it is the starting point. But the same ideas must be used to satisfy all stakeholders.

Elements of Customer Value

When selecting any product or service, customers factor in both its benefits and costs. For example, when you purchase a bottle of wine, both the benefits and the costs come quickly to mind. And then all it takes is a little mental math to decide whether a given bottle is worth it to you.

Therefore:

Customer Value = Benefits − Costs

The total benefits provided by a product or service are called its "worth," which has a financial value. The customer costs of a product or service include, of course, the amount to be paid for it. But there are often other costs as well. In the case of a computer printer, for example, the cost of printer cartridges, paper, and maintenance must be counted. Costs can also include the hassle of changing to a competing product—the transition cost. Some costs may be important to the customer and some not. Thus, like benefits, costs are calculated and weighted in importance only by the customer. A segment of consumers, for example, may discount the cost of printer cartridges when they buy a computer printer, although the printer cartridges may, over time, be the most expensive part of the purchase.

Most products are made up of a basket of features that provide benefits to consumers in a specific market segment. For your own purchases, such as a computer, you evaluate your needs—for instance, word processing, battery life, portability—and look to see which product has features that match your needs. Then, of course, you look at the costs to see if it is "worth" buying.

Through market segmentation, companies attempt to find relatively large clusters of customers willing to pay a premium for certain benefits. You can, for example, buy computers in various market segments, such as those with the most powerful word-processing features, those with built-in wireless capability, and those that cost the least. The same thing is true for cars. If you want high mileage, you might be interested in a hybrid, such as the Honda Civic or the Toyota Prius. If you are interested in prestige you might buy the new, high-performance Cadillac, BMW, Corvette, or Jaguar.

It is important to emphasize that product features become benefits *only* when they address the needs of the consumer. That is, benefits are specifically those features for which customers will pay money. Thus, you must begin by identifying and quantifying the specific customer needs you want to serve.

There are many different kinds of needs, both tangible and intangible. Tangible needs include mobility, communication, and hunger. Intangible needs include styling, prestige, identity, fear, security, love, sex appeal, and group identity, such as being associated with environmentally benign "green products." The broad range of potential customer needs is captured in psychologist Abraham Maslow's hierarchy.[4] Maslow's hierarchy is a five-level pyramid of basic human needs: (1) physiological; (2) safety; (3) love, affection, and belongingness; (4) esteem; and (5) self-actualization. In Maslow's hierarchy, basic physical needs such as breathing and sustenance must be satisfied first. Once these are satisfied, the next level of needs is activated. At the top of his pyramid is the need for self-actualization, which Maslow described as what a person was "born to do." Jascha Heifetz, for example, the greatest violinist of the past hundred years, was born to play the violin. Others might say they were born to drive their Harley-Davidson motorcycle.

Maslow's hierarchy suggests that there are an unlimited number of customer needs that can be addressed. Each higher level opens up families of possible products and services in each market segment, such as food, communication, transportation, and entertainment. In each category, there are also endless ways to balance benefits against costs to create unique products and services.

Consider, for example, the possibilities created by different combinations of quality versus convenience for consumer entertainment and communication products, as suggested in Figure 4.1.[5] It illustrates that whenever a product is significantly improved in either its quality or its convenience, it can spawn a new industry. For example, consider the journey from radio, to black-and-white TV, to color TV, to HDTV. These product innovations significantly increased quality through better sound and images for the customer. Alternatively, the journey from the LP record to the CD player to the iPod represents major innovations in the dimension of customer convenience, in terms of both choice and portability.

A colleague once asked us, "Do you think the success Steve Jobs is having with the Apple iPods is a fluke?" We suggested that it was highly unlikely. For example, the iPod Nano was as far to the right in Figure 4.1 as you could make it. This doesn't happen by accident. In Japan, by contrast, the instinct is to add on every product feature imaginable, to go up the chart. Jobs took all features out but two—audio and choice—in order to minimize size and maximize convenience so he could move to this "white space" without competition. The new video iPod is in keeping with that theme. He is retaining the simple, convenient format and now moving vertically. That doesn't happen by accident either.

Most innovations are combinations of improved quality and convenience features, such as compact disks (CDs) versus long-play (LP) records. Video on demand will soon extend HDTV by adding more convenience to create another industry. Eventually we will have mobile, speech-controlled, head-mounted HDTV video receivers, iHDTV.

The products shown in Figure 4.1 are all major, disruptive innovations. Most innovations are improvements to already-identified products and services. But these smaller innovations can still distinguish a product. Consider, for example, the addition of nonslip rubber pads on toothbrush handles, which allow the customer to securely hold the brush in a wet, slippery hand.

Few companies or individuals think systematically about all the dimensions of a product's or service's possible benefits. Steve Jobs does, and that is why he is so envied. Charts like Figure 4.1 can be developed in any business segment to help visualize potential trade-offs and to find the "white space" where there are unmet customer needs that you can address better than the competition. Try it for your activity to see how you are positioned and whether you have additional opportunities for innovation.

Different people will place different financial values on the benefits a product or service provides. That is how flea markets work: one person's junk is another's priceless find. For a given product or service, you have a number in mind for what its benefits are worth. If you can get the product for nothing and it has some benefits for you, the customer

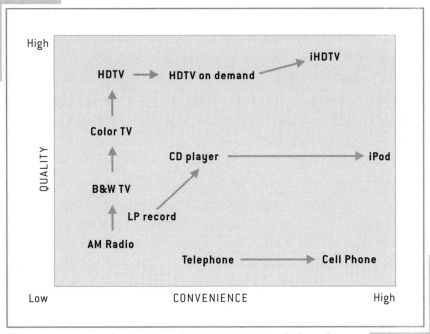

Figure 4.1: An example of different consumer electronics products categorized according to their quality and convenience. Experience has shown that whenever someone creates a new product or service that is a significant improvement in either quality or convenience, it can create a new industry. Simple plots like this for different industries can be used to help identify the "white space" where major new innovations are possible; we call this "white-space innovation."

value to you is the product's full worth. That is why even simple gifts are so nice.

Include All Elements of Customer Value

Providing maximum customer value means considering all factors that can contribute to value, including the customers' *experience* in using a product or service, which is often a major source of customer satisfaction and potential competitive advantage in the exponential economy.

When Toyota developed the Lexus, for example, one key element of customer value was to have dealers provide the highest level of service in the industry. The company did extensive studies in both Europe and the United States to understand its potential customers' needs. Two of the targeted competitors were BMW and Mercedes, which did not have good reputations for customer service. Toyota's research found that less than 50 percent of Mercedes owners used the services offered by their dealers. Customers wanted to talk directly to the people who were going to work on their cars and they weren't able to do that at Mercedes service shops. In response to that customer need, Lexus dealers have white-shirted "diagnostic specialists," mechanics who interact directly with customers. Customers can also relax in comfortable lounges while viewing spotless service bays through large plate-glass windows.[6]

On the other side of the coin is the original videocassette recorder (VCR), designed to record television programs as well as play prerecorded VCR-formatted movies. It was a transformational innovation, but most people had difficulty setting the clock right, let alone programming the VCR to record shows at the scheduled time. These difficulties significantly reduced the value for the customer. Intelligence and education level did not seem to be factors for the more than 60 percent of consumers who could not use their VCRs efficiently.[7] No wonder there has been a rush to the TiVo-type digital recording devices. The consumer experience of TiVo, while still not perfect, is far easier.

Another example is the 3M Post-it easel pads that you may have in your office. Easel pads have been around for decades. 3M experimenters differentiated a commodity by making the product much more convenient to use. First they cut a handle in the cardboard top so you can easily carry easel pads to meetings. Then they put sticky adhesive at the top of each page so you can stick sheets on walls without having to bring an easel. These improvements were simple, but they required thinking about the customers' experience and what they needed.

The makers of throwaway cameras also got the consumer experience right. While the photos you take may not be art, they are acceptable for vacations or weddings where you just want an inexpensive remembrance of a good time. The value of the convenience provided is much more important than the quality. No wonder throwaway cameras often appear on each person's chair at wedding receptions.

If your experience with a product or service doesn't match your expectations, it can actually detract from the product's value. Take the most widely used word-processing program, Microsoft Word. It is loaded with features most users don't even know exist. Even after years of use, most users are taken aback when a colleague says, "Oh, did you know you can use 'document map' to easily view the headings of your paper as you produce it?" You think, "How could they design this so I can't find these things?" Then, there are default features, such as outlining, which have been made so "automatic" that many users can't figure out how to turn them off. With no easy workaround, customers become irate and stop using the features. These items are, after all, supposed to be productivity tools, not sources of more frustration and dissatisfaction.

When you are designing your new product or service, take a comprehensive view of its value. Yes, physical features are important, but so are the users' experiences, and so are the intangible attributes, such as emotion and identity. Remember also that in the exponential economy, customers have a great deal of knowledge about you and your products and services. A good rule is to assume they have complete knowledge. If you produce an excellent product or service, the word will spread at the speed of light over the Internet and you will experience great success, such as happened with the Apple iPod. But if some aspect of your product fails, the world will know that just as quickly.

Family of Higher Benefits

Certain products and brands command a premium because they address a wide array of benefits, including prestige, identity, sex appeal, risk reduction, durability, service, and the customers' overall experience. Consider:

- Starbucks
- Apple
- GM's Corvette
- Gucci
- Mikimoto

Obviously, these companies address needs other than the tangible elements of the product. In the list, you can easily identify what else they are selling. For Starbucks, it is fresh, customized coffee in a comfortable, friendly setting with Internet access. After a few visits they also know your name and the drink you prefer. This allows Starbucks to sell a product for $2 to $4 that can be bought at Dunkin' Donuts for $1. Remarkably, the Starbucks "experience and choice" is worth more than the product. For Apple, it is user-friendly products and an appeal to the "dare to be different" identity of its customers. For Corvette, it is power, fun, and sex appeal. For Gucci, it is prestige. For Mikimoto, it is a guarantee of quality in a market where most consumers cannot judge the excellence of pearls. The guarantee reduces risk and increases value.

Michael Markowitz describes a hierarchy to aid in identifying possible needs and benefits, similar to Maslow's.[8] The hierarchy, shown in Figure 4.2, emphasizes that the best products and brands touch our needs at multiple levels, including the need for "deeper meaning." It is clear that Kodak, for example, knows it is selling benefits at many levels, including "immortality." Kodak's advertising touches all levels of Markowitz's hierarchy to maximize the customer benefits and thus the customer value. In the past, before it had serious competition, it allowed Kodak to sell its print products at a premium.[9]

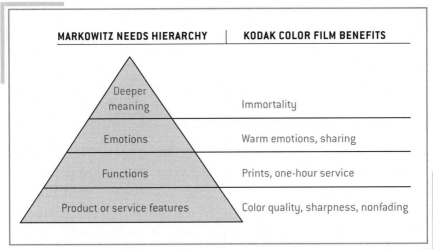

MARKOWITZ NEEDS HIERARCHY	KODAK COLOR FILM BENEFITS
Deeper meaning	Immortality
Emotions	Warm emotions, sharing
Functions	Prints, one-hour service
Product or service features	Color quality, sharpness, nonfading

Figure 4.2: A version of Markowitz's hierarchy to help identify "deeper" needs and thus provide additional customer benefits. Create a Markowitz hierarchy for your product or service. See if you are taking advantage of all the ways you can create value for your customers.

Quantifying Value

Quantifying, or at least estimating, customer value is a subtle exercise. After all, customer value changes over time. One day pink cars are cool and a few years later you can't give them away. If you ask customers to tell you how much something new is worth, they may not fully understand what you mean. And until they put their money down, they have no real motivation to clarify their judgments.

All these difficulties, and more, exist when you attempt to quantify value. But you can at least estimate customer value. The tool we describe below will help you focus on your customers, give you additional insights, and make you more sensitive to the multiple dimensions that make up customer value.

Earlier in her career, Anne had been a marketing director for a large, well-known corporation. It was once a great company but now is out of business, and we were curious about why. The CEO and senior management, Anne told me,* were consumed with increasing the stock

*This section is in the voice of coauthor Curtis R. Carlson.

price. Every public company wants a higher stock price, but increases have to be based on a solid foundation of customer value. Unfortunately, management did unproductive things like buying back the company's stock. Yes, the price went up as long as the corporation bought back the stock, but as soon as it stopped, the price went down again. Buying your own stock at a premium price does not create great products, which is the only way to increase the real value of a company. And, as Anne said, "After a few silly moves like the stock buy-back plan, the company got in trouble because it didn't have the great products it needed. It went away."

Then Anne spoke with passion about why understanding customer value is the key to success. Not surprisingly, few at her former corporation were interested. It was a company of executives who felt that they knew what customers needed. Besides, they said, "Everyone knows that you can't understand or measure customer value. Why bother talking to customers?"

One of our collegues, Len Polizzotto, shares our passion about focusing on customer value. He described to us Value Factor Analysis, a tool he had developed that helped estimate the customer value of a product or service when compared with others. He had developed it as a university professor teaching engineering students how to think about a new product's value relative to the competition and alternatives. It also taught the students how to think about the trade-offs that must be made when designing any new product or service. To do this you need a way to understand which attributes and costs matter most to your customers. After describing this tool, he laughed ironically and said, "It couldn't hurt." What he meant was that if you wanted to understand your customers, tools like this are essential. We teach Polizzotto's Value Factor Analysis to our employees and partners at SRI.

The buy-or-not-buy decision by customers is made on the basis of the product's total benefits versus the money that has to be spent: that is, its worth compared with its costs. Consumers usually think about a product or service in terms of its *comparative value,* and make a buying decision based on it. Consequently, Value Factor Analysis alters the formula for value we gave above by making a relative comparison, which is a perceptual quantity. When you divide benefits by costs, it sharpens the relief between two or more products.[10] Thus:

Value Factor = Benefits / Costs

Because this expression is more intuitive, it is used in the definition of a value proposition, as described in Chapter 5. As we will now show, it also allows you to compare the relative values of different products or services.

As indicated in Figure 4.1 and as we will discuss in the Appendix, the benefits of a product or service can be further divided into two separate dimensions—quality and convenience. Similarly costs can be subdivided into those representing quality and convenience. In the example given below, however, we simplify the analysis in order to focus on the main conclusions.

Figure 4.3 shows the Value Factor. Our colleague, Herman, is an environmentally conscious ("green") consumer. He already has two electric cars and he recently bought a Toyota Prius. Why did he buy a Prius and not a Ford Taurus, which is about the same price? As we will show, the Prius has significantly greater customer value for Herman than does the Ford. Conversely, for other people the Ford provides greater customer value.

The top left column lists a few of the many features or attributes an automobile can have. Next to it are estimates of the importance of these attributes to Herman, on a scale from 0 (no importance) to 5 (very important). For example, pollution from carbon dioxide is very important, so it gets a 5. A rear fancy "spoiler" has no importance for Herman, so it gets a 0.

The third column shows how well a Toyota Prius "satisfies" these attributes, also on a 0 (does not satisfy) to 5 scale (satisfies superbly). Since the Prius gets up to 60 miles per gallon, it does an excellent job of satisfying the criterion for pollution and thus rates a 5.

The last column shows the benefits resulting from the different attributes, computed by multiplying the two scores for "importance" and "satisfaction" together. The sum of all the benefits is a measure of the overall benefits of the Prius, or its worth. The objective is to make this number as big as possible to provide maximum customer benefits.

Next we compute the total costs for a Prius. We want to keep costs that are important to the customer as low as possible. The bottom half of the chart shows a calculation for the customer's costs. Again the "importance" scale goes from 0 (Herman is completely cost-insensitive) to 5 (high cost is a big disadvantage). Under "expense," we show dollar signs that go from none to five. No dollar sign means that feature of the Prius is very inexpensive; five dollar signs means it is very expensive. Since the cost of gas is only moderately important to Herman, the Prius gets a 3 for "importance" and, since it gets excellent gas mileage, it gets only one dollar sign for "expense."

"Cost" for each entry is estimated by multiplying its "importance,"

PRIUS			
QUALITY	IMPORTANCE: 0–5	SATISFACTION: 0–5	BENEFIT: 0–25
Pollution	5	5	25
Styling	2	2	4
Reliability & durability	5	4	20
"Green" identity	5	5	25
Rear "spoiler"	0	1	0
TOTAL BENEFITS ("WORTH")			74

COST ATTRIBUTES	IMPORTANCE: 0–5	EXPENSE: 0–$$$$$	COST: 0–25
Base price	1	$$$$	4
Gas costs	3	$	3
Repairs	4	$$	8
Insurance	2	$$	4
TOTAL COSTS			19

VALUE FACTOR FOR A PRIUS (Sum of Benefits/Sum of Costs)	3.6
VALUE FACTOR FOR A FORD TAURUS	1.4

Figure 4.3: A Value Factor Analysis chart showing how to estimate the comparative value of different products or services, using the Prius hybrid automobile as the example and the criteria of an environmentally conscious consumer, Herman. A second analysis, also using Herman's criteria, was performed for a Ford Taurus. The Value Factor was 1.4. Since 3.9 is significantly larger than 1.4, it indicates that a "green" customer like Herman would prefer the Prius.

as determined by Herman, multiplied by its "expense," as represented by the number of dollar signs. All the costs are then added to give a measure of the Prius's total costs. The total costs are finally divided into the total benefits to compute the Value Factor, which has a value of 3.9, as shown in the figure.

A similar calculation for a Ford Taurus using the same attributes and costs gives a value of 1.4 using Herman's criteria. This simple calculation helps indicate why Herman bought the Prius and would not even test-drive the Ford. Other people may value styling and other features that are represented by the Ford, such as a fancy spoiler. For these customers, the score for the Taurus may be higher than the score for the Prius. *Vive la différence.*

The calculation shown in Figure 4.3 is highly simplified, but it does provide a straightforward way of identifying the attributes that are significant to a customer. When talking with Herman about these results, we pointed out that the Prius was much more expensive than a comparable conventional car and that it would take more than a decade to get this money back in gas savings.

Herman responded, "Why do people buy BMWs? What's the 'payback' from an Acura or Lexus or Cadillac, rather than the near equivalent but cheaper Honda, Toyota, or Chevy product? To some of us, owning a hybrid is cooler than owning a BMW, which we could equally afford." Clearly, different people have different yardsticks to measure value.

We always find it beneficial to go through Value Factor Analysis to help identify all the possible features we should be considering for a new product or service. It also helps us frame questions for our prospective customers about how they perceive the benefits from different features. If there was doubt about the conclusions from Value Factor Analysis, additional market studies could be performed to obtain better, more quantitative information. Those in marketing attempt to use these kinds of tools to find clusters of benefits that appeal to large untapped market segments, which is called "white-space innovation."

Consider another everyday example: You go through the same process of assessing worth when selecting a hotel. One interesting study of more than 400 respondents showed that the decision to pick a hotel is determined by the astonishing total of 1,275 hotel attributes. Fortunately for the hotels, each traveler has only three or four attributes that are the basis for comparative decisions. This example is like Herman and his selection of the Prius. For him just a few characteristics determine his decision. Having a fancy spoiler is not one of them.

There are other techniques that can be used to help quantify customer value. We highly recommend the books of Harry Cook to those who want to learn a broad array of helpful ideas.[11] His methods for quantifying customer value can be very useful, but they do require some additional work. You should read his publications and those of others, and then decide when these more formal methods are warranted. At a minimum, always complete a Value Factor Analysis and iterate it with your teammates and prospective customers. It will open up your eyes to the possibilities available to you.

DISCIPLINE

2

value creation

CHAPTER 5

it's as simple as naBc:

HOW LIZ got HeR BIG JOB

IF YOU CAN'T STATE YOUR VALUE PROPOSITION,
YOU DON'T UNDERSTAND YOUR JOB.

An Opportunity

Liz was the editor of nonfiction books at a top publishing house when the job of editorial director opened up. She saw it as a great opportunity, but she was worried. Liz is smart, savvy, and energetic, but normally the job would go to someone in his or her thirties or forties with ten or more years of experience. Liz was twenty-seven and looked six years younger. She thought she had the intelligence and skills to do the job, if those making the ultimate decision could look beyond her youthful appearance. But would her application be taken seriously?

Liz called her father, Art, to get some advice about what she might do. Art asked her for her value proposition. "My what?" she asked. He explained that every business transaction has at least two value propositions. In this case, one would be from the hiring manager at the organization to the job candidates, explaining why they should take the job. Another would be from her to the company, explaining why she was the best person for the job. Each side needs to persuade the other.

Remarkably, many people go into interviews without thinking of how they can add value. Some sit there expecting to be "sold," while others cite items on their résumé. They miss the opportunity to fully describe why they will be the best person for the enterprise. Thus, Liz could distinguish herself by using her knowledge and packaging it into a compelling value proposition: her vision of how the department could create more value for its customers and the publishing house. And it would be in a form that would facilitate a productive conversation about the job.

Art acted as Liz's partner as she developed and improved her value proposition. They iterated it a dozen times. She talked to many others, continually synthesizing into her presentation the new ideas she uncovered. In the end, when she came to the interview she addressed four issues:

- The Publisher's *Need:* The nonfiction department's overall book list was not making money, although a few books on Liz's list were. Its normal customer base of libraries was purchasing 5 percent fewer books every year, setting a terrible downward cycle into motion. The department needed to keep its brand at the highest quality, yet it had authors asking for royalties that were too high to leave any profit. So it needed to maintain its brand and quality, expand its customer base, change its book portfolio to be more profitable, and somehow pay less in royalties.

- Liz's *Approach:* She proposed to change the mix of books based on an analysis of what was profitable, since 90 percent of the department's profit came from 20 percent of the books sold. She proposed new models to expand the customer base, such as introducing lower-priced books, with subject matter that was high-quality and would appeal to a broader readership. She suggested ways of enlisting more emerging authors in addition to the established ones, thereby lowering royalties. She also described new ways to combine previously published material into new books.

- *Benefits per Costs* for the Publisher: Liz produced a spreadsheet of the current revenue and earnings, and she forecast 10 percent per year revenue growth and 15 percent profit based on her suggestions. She also showed how the publisher's brand would be enhanced, causing a "virtuous cycle," attracting more authors and therefore more books and more profit as well. Her approach would not cost more than the publisher's current approach.

- The Publisher's *Alternatives:* Doing business as usual was causing a slow and inevitable decline. Her approach was the only one presented that offered the publishing house a way to revamp its approach and grow. She was also the only candidate who quantified, in a realistic way, the financial advantages of the changes she suggested.

Partly because of her value proposition, Liz was unanimously chosen as the top candidate and given the position. In her first yearly review, her manager gave her a double-digit raise and said that she exceeded all targets and was a "blessing" to the publishing house. When we talked to Liz she said, "There was no way I could have won this job without creating a value proposition. The format was perfect. You start off reviewing the company's needs, and they agree with you where you get it right. You can then have a deeper discussion with them about their needs and learn more before you go to your approach. It was perfect, just perfect."

If you are like Liz, or most of us, when you start a new project it seems impossible. It is scary. We are often asked, "How do you take a vague, incoherent idea and turn it into one that creates compelling new customer value?"

The process Liz used provides the answer. It has been developed, tested, distilled, and refined for years by us and our colleagues. It is the origin of hundreds of projects, more than fifteen new companies, and two broadcasting Emmys, including one for developing the U.S. standard for digital high-definition television.

As you read this chapter some of you may say that you already do these things. Yes, you probably do some of them. Because of the effectiveness of this process, many other creative professions, such as music teaching, moviemaking, journalism, and public relations use variations of these ideas.[1] The celebrated dancer and choreographer Twyla Tharp wrote a book called *The Creative Habit.*[2] Her view is that achievement—

whether in dance, business, or restaurant cuisine—is the result of preparation, effort, and proven practices, all disciplines that can be learned. She emphasizes closely observing the world and starting by getting your ideas down on paper. As you read this chapter, you will think of your own examples about how these ideas apply to other activities.

Even though these ideas are not complicated, very few people and even fewer enterprises systematically apply them. After learning about and meeting with senior managers from hundreds of organizations, we have found only a handful that do. These few companies tend to be leaders in their fields, like GE, P&G, Baldor, Toyota, 3M, and W. L. Gore.

Value Propositions: NABC

Every important innovation opportunity requires a value proposition. When it's missing, the result is confusion, poor communication between employees and management, a lack of focus on the customers' actual needs, and wasted organizational resources.

The goal of every innovation is to create and deliver customer value that is clearly greater than the competition's. The difference has to be compelling to the customer, which is expressed as:

> "Our New Customer Value is *much greater* than the
> Competition's Value."

But how do you, like Liz, start to develop new value for the customer? Begin with your value proposition. It is the nucleus for value creation, because it addresses the four fundamental questions that must *always* be answered when you start to create new customer value. These questions are:

1. What is the important customer and market *Need*?
2. What is the unique *Approach* for addressing this need?
3. What are the specific *Benefits per costs* that result from this approach?
4. How are these benefits per costs superior to the *Competition's and the alternatives*?

In an essential way, these ideas are not unfamiliar. You create value propositions as a matter of course in everyday life. For example, if you have a visitor and it is time for lunch, a typical conversation might be, "June, I understand you are as hungry as I am [Need]. Let's go have lunch at the Company Café [Approach], instead of McDonald's [Competition] because, for the cost of McDonald's, it has great food, it's quiet, and we can continue our conversation [Benefits per costs]."

Who determines the worth of this value proposition? Do you as the host? No, it is your visitor, June. And June might say, "Thank you, but I promised my children that we would go to McDonald's today, so I would prefer going there. My children will be happy playing on the slide, and we will be better able to continue our conversation."

This simple example demonstrates several other important points. First, the order of the questions—need, approach, benefits per costs, and competition—does not matter. What is important is answering all of them. Second, there are many different approaches for addressing a particular need. Your goal is to develop an approach that provides superior customer value—benefits per costs—when compared with the competition and alternatives. To be effective, value propositions must be quantitative and easily understood. A picture, image, or mock-up greatly facilitates communication of your ideas. When it's appropriate, always create one.

Value propositions are difficult to develop in business, because initially you don't know enough about any of the four questions. In addition, these four key questions interact with one another. One must repeatedly improve or "iterate" value propositions by getting feedback from others to make them complete and compelling. As we discussed in Chapter 2, you must repeatedly go back and forth between the unmet needs of your customers and your sources of new ideas to create new innovations. Without a process to help, answering the four questions in a value proposition tends to be either elusive or impossible.

One reason value propositions are hard to develop is that we all love talking about our "approach," to the exclusion of the other components. We have listened to many hundreds of business presentations during our careers, and in the beginning they always sound like this:

Need, **Approach,** Benefits per costs, Competition.

They are all about "approach." For example:

> "What the world *needs* is a little red wagon.
> Our *approach* is to build a little red wagon.
> The *benefits* are that we will have a little red wagon.
> There is no *competition,* since our wagon is a nice shade of red."

It's all approach, with no need, benefits per costs, or competition. You might think that this example is an exaggeration, but it isn't. People are always consumed with *their* approach—after all, they thought it up. They want to tell you about it and they want you to appreciate it. If you don't, they often assume it is your problem, not theirs.

When you begin developing your value proposition, it should initially look more like this:

Need, Approach, Benefits per costs, **Competition**

In this case the focus is on understanding the customers and the potential competition. That is, it is all about understanding the market ecosystem—the customer and market needs and the current and potential future players. Over time, you can create an approach, or refine an existing one, with compelling benefits per costs when compared with those of the competition.

When it comes to innovation, most enterprises are a Tower of Babel. The staff and decision makers literally speak a different language. The staff will initially talk about "nAbc," because they are focused on their approach. There is no need, benefits per costs, and competition. But the decision makers will be initially interested in "NabC," because they want to address their customers' unmet needs and beat the competition. They assume the staff can figure out the approach and benefits per costs once the needs and competition are understood. With no process in place to put all the pieces together, the staff and decision makers are at loggerheads. The staff will say things like "The decision makers don't get it" or "They don't support us." When you talk to the decision makers they also will say things like "The staff doesn't get it" or "We're unable to get the staff to do useful things." Remarkably, both groups say the same things about the other and both feel powerless. The solution is to have a value-creation process in place where everyone speaks the language of customer value, including all elements of a value proposition, NABC.

Remember Frank from Chapter 1. When he came into the room he brought a "nAbc" presentation. He was frustrated because no one would take him seriously. Actually that wasn't the problem—no one could understand him. In most organizations this is a common problem. If you are a decision maker with no value-creation process in place, you too will probably be befuddled when a Frank walks in, because you don't have the time to figure out what he is saying in order to determine whether it has value for the enterprise. At that point you have two poor choices: 1) You can politely ignore him or 2) you can send him off, saying "Do more homework" in the hope that something will happen to help you make sense out of what you just heard. These choices demoralize the staff and leave the decision maker feeling ineffective. Absent a way to engage the staff member, they may eventually walk out the door. Steve Wozniak, cofounder of Apple with Steve Jobs, left HP, a company he loved, because he couldn't get management to pay attention to the potential of personal computers. This outcome is not unique. Hundreds of talented employees, carrying billions of dollars of new value in their heads, leave companies they love each year. As a champion, you must make sure value-creation processes are in place to engage your Wozniaks in order to capture their important innovations.

The next time someone gives you a presentation, keep the four ingredients—need, approach, benefits per costs, and competition—in mind and see if they are all included. If they are, you are talking to a very rare and special person. But if they aren't, be tolerant—developing a good value proposition is extremely difficult. That is why it must be part of a thoughtful improvement process with your team and organization, not an ad hoc activity.

Below are some short, effective value propositions. The first one is from Paul Cook, a Silicon Valley Hall of Fame entrepreneur. It is the value proposition presented to a cable company executive for a video-on-demand system. We have labeled the four key ingredients—need, approach, benefits per costs, and competition—so you can see how they all fit together.

———{ VIDEO-ON-DEMAND VALUE PROPOSITION }———

"I understand that you are looking to expand your business. I think we might be able to help.

- **(Need)** Movie rentals represent a $5 billion business opportunity that you currently cannot access. The only parts of rentals that people really dislike are the obligation to return the tapes plus the late fees. Customers find that it is inconvenient and wastes time.

- **(Approach)** We have developed a system that allows you to provide videos on demand to your customers using your cable system, with access to all the movies of Blockbuster. Our approach makes use of one of your currently unused channels, with no changes to your system. In addition, you do not need to invest any capital. Each movie costs your customers $6.99, the same cost as a rental at a video store.

- **(Benefits per costs)** You will receive $5 of new revenue per movie rented, with a margin of 20 percent after paying for the movie costs. Your customers will have all the pause and fast forward functions of a VCR when watching the movie, and they do not have to return the movie when done. Late fees are gone. We estimate you could capture a market share of 20 percent.

- **(Competition)** Our system is patented, and it is the only one to include all of these features. Online rentals represent new competition for both you and us, but they have a handling-cost disadvantage of 75 cents per tape. Sending videos back is inconvenient, plus they cannot provide spontaneous purchases.

Would you like a follow-up meeting to see how we could help you increase your revenue and profit?"

Writing down your initial answers to these questions is the beginning of a proposal or innovation plan for your project. The power of the NABC value proposition is that it is concise and to the point. As we will describe in Chapter 8, it is also the basis for your Elevator Pitch, which is a short, pithy summary of your value proposition; and then, in Chapter 9, we describe your full innovation plan.

Note that earlier we wrote competition *and* alternatives. In many situations there is direct competition, such as Corvette versus BMW. But in other cases there are alternatives that we might not consider at first. For example, for those interested in an exciting driving experience, the comparison might be between a sports car and a motorcycle. In this case motorcycles are a serious alternative with sales growing rapidly.

When someone comes up with a new idea, ask the person to answer these four questions: need, approach, benefits per costs, and competition and alternatives. Communication about an innovation is facilitated through a *common language,* including value propositions. And if the value proposition is not persuasive, the simple NABC template enables you to develop a compelling, persuasive one that will help sell your innovation to your peers, supervisors, and, eventually, customers.

Multiple Value Propositions Are Required

In every situation, *at least* two value propositions are required. The first is to your prospective customers, the people who might buy your product or service. The metrics for them are benefits per costs and how those benefits per costs compare with those of the competition and alternatives.

The second value proposition is to the investors. In an established organization, that might be the manager of the company, the principal of a school, or a contract officer at a government agency. In other cases, it might be to outside investors, such as venture capitalists. In that case, the metrics for success include market size, profit, revenue growth, and return on investment (ROI).

In most situations, more than two value propositions will be required, because you may have business partners in addition to your customers and investors. In the Paul Cook video-on-demand example, the owner of a cable distribution system was a necessary partner in addition to Paul's investors. The value proposition to the cable owner emphasized revenue and profit growth for his company. But you will also see that this brief value proposition included benefits per costs for the cable owner's customers, such as the benefit that at the same price, there is no need to return rented tapes. The cable owner would only be interested if he was sure his customers would be getting additional value too. Typically, in any new initiative, value propositions must be developed for suppliers, distributors, staff, and other partners, if they are all to join the initiative.

Below is an early draft of a hands-free car phone value proposition. It is in outline form, which is the best way to get started. Note that it includes benefits per costs to both customers and investors.

─────{ HANDS-FREE CAR PHONE VALUE PROPOSITION }─────

NEED

- Cell phones are difficult and dangerous to use when you are driving.
- There are more than 500 million cell phones in use around the world.
- Because of the driving risks, many U.S. states and other foreign governments are legislating against the use of cell phones by drivers of moving cars, which would limit cell phone usage in cars.
- Consumers want to continue to be able to use their phones while driving.

APPROACH

- Use voice-activated dialing with a headset.
- Provide additional software for existing phones.

- Make the software "downloadable" to existing phones with a $10-per-year subscription for the "in-car service."

CUSTOMER BENEFITS PER COSTS

- Convenience
  - *Allows increased phone usage.*
  - *Safe, comfortable, and easy to use: does not require a new phone.*
- Quality
  - *Excellent speech recognition for voice-activated dialing: 99 percent accuracy with untrained users.*
  - *Supports twelve different languages.*
  - *Robust performance in noisy environments—better than human performance.*
- New applications: the speech interface allows access to the Internet and other services.
- Cost: $10 per phone per year.

INVESTOR BENEFITS PER COSTS

- New product = increased sales.
  - *Assume our available market share is 10 percent of the 500 million total market.*
  - *At $10 each, then 50 million × $10 = $500 million per year total revenue potential.*
  - *Initial investment needed: $5,000,000*
  - *After three years, achieve revenue of $50 million per year with a return on investment of 5:1.*
- Other sources of revenue are available, because of the speech interface.
  - *Initial discussions under way with other service providers.*
  - *Applications include: navigation, auto service, food.*
  - *Business model TBD.*
- Hands-free auto use can reduce litigation, which many be another revenue opportunity.
  - *Today the average cell phone car lawsuit is around $50,000.*
  - *Opportunity for reduced auto insurance to the consumer*
  - *Business model TBD.*
- Low product risk = prototype developed and demonstrated.

COMPETITION AND ALTERNATIVES

- Existing phones, which must be used outside the car.
- Speech-activated phones built into the car.
  - *More expensive at more than $100 per car.*
  - *Less convenient for the consumer who wants to use a regular cell phone.*
- Possible competitors: Intel, IBM, and Microsoft.
  - *Our demonstrated 10 percent better speech quality and car noise insensitivity enables this application.*
  - *Intellectual property protection: we are protected by a family of fourteen patents.*

This early draft was written after just a few hours of thought. It left many questions unanswered, and some of the early assumptions turned out to be incorrect. But that didn't matter. It was a start, and this initial draft convinced us that the opportunity was important enough to pursue. Therefore we began the iteration process to improve the value proposition and determine whether we could create a compelling business opportunity.

Specific, Quantitative, and Illustrative

Good value propositions are specific, quantitative, and illustrative; they tell a story. When people present their first value proposition to us, we usually give them one to four minutes to present it. We have listened to hundreds of these presentations, and only a few have been passable. Many are unintelligible. But we are sympathetic; we have never gotten one right the first time either. Even with the NABC template in your mind, it takes numerous rounds of improvements to be effective.

For example, the hands-free-phone value proposition says, "Many U.S. states and other foreign governments are legislating against the use of cell phones by drivers in moving cars." This is not good enough. We need to know the dates when the new laws will come into effect and exactly what they will mandate. We also assume that 10 percent of the 500 million phones will want our service. We need much more specific evidence about the number of potential customers and how we will engage them.

Try this simple experiment. Imagine that someone just came into your office and asked for the value proposition for your most important project. What would you say? Try it.

N: My customer's needs are . . .

A: My approach to satisfy that need is . . .

B: The benefits per costs of my approach are . . .

C: My benefits per costs are superior to the competition and alternatives because . . .

Your value proposition template gives you a more powerful worldview, which also applies to everyday business tasks. Recently, we were with a colleague preparing a presentation. He is an expert in understanding customer value, but he was like everyone else in that the first time he practiced his presentation it was about him. That is, he talked about the topics with which he felt comfortable—his approach. We all do this, but presentations like this just exasperate your audience. We asked him to put on his "customer value hat" for each slide he showed. That is, to think about what the audience needed, why the points on the slide represented value for them, and why this value was superior to the alternatives. When he gave the final presentation, he got rave reviews, and a few months later he won a huge contract.

Finally, where possible, you should create a detailed picture of the product or service in use, so the concept of the innovation becomes evident. Below are some tips for creating compelling value propositions.

- Customers
  - *Talk to and interact with your prospective customers: Deeply understand your market space.*
  - *Create a first Value Factor Analysis to make sure you are considering all elements of potential value, from tangible features, to convenience, to the overall experience, and to intangible issues, such as security and identity.*

- Create a prototype, if possible, or at least a picture or mock-up of your product or service.
  - Watch and study your prospective customers using your prototype.
- Competition
  - Study your competition: know the competitors by name.
  - Understand all the alternatives.
- Look to the future: anticipate new competitors and market disruptions.
- Be quantitative: if you are not sure, take a SWAG (Scientific Wild A—Guess).
- Iterate.
  - Fail often to succeed early.
  - Form a "Watering Hole" (Chapter 6).

Innovation Plans

Value propositions are not complete innovation plans. But they address the *fundamentals* of a plan. It doesn't make sense to develop a more extensive plan until these basic questions are answered in a clear, concise, and compelling way. We use the term *innovation plan* rather than business plan, because this phrase emphasizes the urgent, dynamic imperatives of value creation.

In many situations, a short value proposition is sufficient information to decide what to do. But where substantial resources are to be invested, the value proposition must eventually gather and include all the components of a full plan, whether it is a government proposal or a new venture. In Chapter 9 we will show how value propositions naturally evolve into full-blown innovation plans, and other ingredients and tips for success will be described.

All Activities Need a Value Proposition

A value proposition should be developed for all your organization's innovation activities, whether incremental or transformational. That includes the formation of new products and businesses, product improvements, management productivity tools, government policies, education reforms, basic scientific research, and enterprise-wide programs from the finance and human resources departments. It's also an invaluable tool for your personal career development, as we showed in Liz's story.

Steve Obsitnik was a founder of Discern Communications, an SRI International company. Steve learned about NABC value propositions and value creation during the formation of Discern at SRI. Several years later, he said how much he appreciated learning about these concepts, which he uses all the time. He told us that his talks to employees were in the NABC format, since it helped him focus on what his audience needed. He even does performance reviews using NABCs. He gives his value proposition to each employee, and the employee gives one to him. He said it was a remarkably efficient way to develop goals and achieve organizational alignment.

Innovations can be incremental (the development of a new website) or transformational (the creation of a new drug to control diabetes). In most organizations, the value creation process is a series of ad hoc, inefficient steps. But innovations never occur out of whole cloth. Innovations must be developed in a value-creation process. The value proposition is your most important value-creation tool. It allows you and the entire organization to focus on answering the core questions for each initiative. It also provides a tool to allow the compounding of ideas, one upon the other. Without such a tool, exponential improvement, as described in Chapter 2, is not possible.

Discover the power that Liz used to win her new job. The next time you give a presentation, put it into the NABC format. Make it as specific and quantitative as you can. Add a picture or sketch of your approach, if appropriate. Then find a partner, and others, and ask them to critique your content and presentation multiple times. We'll bet that when you give your presentation, your colleagues will notice a positive difference. Your value proposition is the starting point. Eventually, it will become the core of your Elevator Pitch and your complete innovation plan.

These ideas have been used at SRI International for all organizational activities, including presentations to customers, formation of new companies, requests for capital equipment, and changes in human resources policies. Since every activity within an organization has a customer and a need for resources, every new activity requires a value proposition to demonstrate why it is the best use of the organization's time and money.

watering holes for creating value:
the day the BBC walked in

THE OBJECTIVE IS TO BUILD NEW CUSTOMER
VALUE AT RAPID, EXPONENTIAL RATES.

A Wake-Up Call

One Monday at 7:30 A.M. not too long ago, senior managers from the British Broadcasting Corporation (the BBC) came to participate in the SRI Discipline of Innovation workshop. They are an outgoing, articulate, and witty group. The team members were full of energy but they had little idea what was in store for them.

The BBC was formed in 1922 and is considered to be the most respected news and entertainment organization in the world. Its roots go back to the early days of radio and television. It broadcasts not only in English but in forty-two other languages in Asia, Europe, the Middle East, the Americas, and Africa. Not only does the BBC provide international news coverage, it has also been a pioneer of top-quality entertainment programs, such as *Masterpiece Theatre,* as well as shows that have been adapted for commercialization in America. Some examples include *Who Wants to Be a Millionaire?, The Weakest Link,* and *Whose Line Is It Anyway?*

While the BBC is often seen by outsiders as a superbly creative enterprise, its management did not think it was keeping up with worldwide changes in media and broadcasting. The management team was searching for a partner to help invigorate the "creativity" of the organization. They signed up for the SRI Discipline of Innovation workshop, and over a period of a year and a half they brought six waves of senior executives to the workshops in Silicon Valley. At SRI we were excited to find a quasi-governmental organization wanting to improve its innovative practices, especially given the enormously high regard we held for the BBC.

We noted earlier that most people cannot concisely state their value proposition. To test this, when the BBC managers first arrived we immediately took them, one at a time, into four separate rooms that had video cameras. Once there, with the door closed, we said, "You have just gotten on the elevator with the director of the BBC. He turns to you and asks, 'What important project are you working on today?' You have ten seconds to think about this, and then we will turn on the camera for exactly one minute. At the end of that minute a bell will ring and we will stop the tape." It was a wake-up call that the workshop was going to be intense. Luckily, no one expired, and all were able to get through their one-minute Elevator Pitch. Since so many of them had television experience, it was fun to see how articulate they were and how well they made eye contact with the camera.

We then brought the entire team together for a meaningful discussion about the exponential economy and the speed of change in broadcasting around the world. With our workshop instructors and facilitators, Herman Gyr and Laszlo Gyorffy, guiding them, they mapped the changes in their world. As they created their "world map" of new customers, potential competitors, and transformational technologies, they began to get a glimpse of the enormous opportunities—and threats—opening up in their world.

The next question was "How does one respond to these opportunities and threats and turn them into sustainable innovations?" With new digital devices in their hands, customers are less inclined to passively sit and watch TV or listen to the radio. They are continuously communicating with one another, playing electronic games, and surfing the Internet, which decreases the time spent with traditional broadcasters, such as the BBC. We noted that the only way one can meet these challenges is to create more customer value with innovative new approaches that leverage these emerging developments.

We then introduced the idea of a "Watering Hole," a multidisciplinary, collaborative environment used at SRI where participants come together to *improve their value propositions* and create more customer value. In an SRI Watering Hole, Elevator Pitches and innovation plans are presented and participants give feedback on how to make the value propositions more accurate, crisp, and comprehensive. We explained that there needed to be such forums in the BBC, where the hard innovation topics could be tackled.

The BBC workshop participants were divided into four teams. Each tackled an innovation opportunity by developing an Elevator Pitch with the instruction to present it in three minutes. The groups went to work, amid comments like "We need an hour, not three minutes. Are you sure this is possible?" They were used to hour-long presentations back home, where program developers would prepare for weeks or months and then present to a "controller," who would give a yes-or-no decision. Typically, only one program out of ten or more would be funded. We were asking them to play a different, supportive role with one another to improve each team's value proposition.

Value propositions improve rapidly only if you get new ideas and useful feedback. We had noticed early on that the wry, acerbic British wit and tendency to debate would jump out even when people tried to be supportive, so we divided the audience into "green hats" and "red hats." Those designated as green hats could only give positive feedback, while the red hats would suggest ways to improve the pitch.[1] At first, they all wanted to be red hats, since they were so practiced at it! But we insisted that everyone had to play both roles at different times as the Watering Hole progressed. If you were appointed to be a green hat, you could only point out the positive elements of the presentation, such as "I appreciate how effectively you quantified the market need," and "I like how you painted a clear picture of the audience demographics." When you go into a Watering Hole as a presenter, you quickly realize how important it is to get this feedback to help you understand what worked. Positive feedback also encourages you to work even harder to get ready for the next presentation. We even worked with the red hats so that they could frame their comments in a helpful way. So rather than saying something like "That's the most muddled mess I have ever heard," they would say, "This would be more effective if it were quantitative. You might consider saying our new product will reach 25 percent more audience members in the twelve-to-eighteen age range." Both red hats and green hats were asked to be brief, specific, and quantitative in their suggestions.

Another crucial behavior surfaced. Because they were so eager to debate, we would make sure the presenters didn't immediately respond to the feedback—they stood there in front of the entire group and listened intently. As the urge to debate with the red hats emerged, they would just thank the feedback providers. To make sure the comments were not lost, each presenter was assigned a "coach." If the presenter survived the feedback, she or he would meet offline with the coach and review the feedback.

Midway through the week, after we had explained value propositions, the NABC format, and had gone through a Watering Hole experience, we showed the videotapes of their one-minute presentations from the first morning. Not surprisingly, the one-minute pitches were not exactly Academy Award quality. We have sat through hundreds of similar pitches and *no one* ever gets it right the first time—not us, not the BBC professionals, no one. When people are asked to describe the new customer value they are developing, it is extraordinarily difficult to specify.

After multiple Watering Holes, with each team revising its NABC again and again, we came to the final day, when teams had five minutes to present complete innovation plans to the group. Each team got rousing ovations in the final round. Their presentations included cardboard cutouts, funny skits, pictures, and artwork that helped us all visualize the innovative programs they were proposing. Everyone agreed they were among the best presentations they had seen.

Epiphanies

Five epiphanies emerged from the BBC's Watering Hole experiences. The first was the realization that they did not really understand their customers. They talked about them, but they didn't really know them. They would use language like "We need to fill the 8 o'clock slot" or "We need a comedy at 11." As one response to this realization, they changed NABC to aNABC, where "aN" reinforces the focus on "audience need" in addition to approach, benefits per costs, and competition and alternatives. Subsequent programs such as *Fat Nation* benefited from this realization. They talked to audience members and found out about people's real concerns.

The second epiphany was that disciplined processes do not necessarily squash creativity, which was one of their major concerns. Rather, they discovered that the processes we described were actually *creativity amplifiers*. Specifically, Watering Holes provided a forum to gather staff from across the BBC to gain needed information and expertise to improve proposals faster. Previously, there was no good way to do this.

Third was the unanimous agreement that important innovations require hard work. Only with a systematic process that helped answer essential questions at rates matching those of the exponential changes in their environment could they be successful. As a result, with the help of our colleague Herman Gyr, they adopted the Watering Hole practice at home under the leadership of Caroline Van Den Brul, who served as Group Leader of "Making It Happen," which was the organization's change program. They then went further and used a team of volunteer facilitators to conduct Watering Holes.

The fourth epiphany was that their larger goal was not to be the "most creative" organization in the world but, rather, to deliver the "greatest public value." In other words, to employ innovative approaches that serve the needs of their various audience segments. They realized that creativity is just one input that needs to be at the service of the end goal—creating customer value. The focus on customer value for the public was a profound realization that has the potential to redefine how public organizations conceive and execute their missions. It is now part of the new charter for the BBC. The BBC is engaged in a process to understand how to measure and deliver programs that realize this goal.

The fifth epiphany was that the BBC's business model had to change. The BBC realized that their customers were being segmented into different markets by the emerging technologies and business models of the exponential economy. The BBC, rather than being a "broadcaster" of content to a passive audience, had to become a "platform" that used multiple channels, including games and interactive Web services, and allowed audience involvement in creating content. As one person said, "Imagine if every school had cameras and teams of teenagers creating programs. They would come up with ideas we would never dream of and tell us where our customers are going with their interests."

One BBC participant, Andy Parfitt, controller for Radio 1, started his new innovation processes at home by sending out all his employees to take pictures of their audience. Large blowup pictures of their audience now adorn the walls, staircases, and halls of their workplace. Talk about customer focus! Parfitt said, "In the twelve months since my visit to SRI, we have totally revolutionised the creative processes at Radio 1. It is clear that ideas we have generated, such as *Star Pupil* and our approach to *Glastonbury* and the *Big Weekend,* have been step changes in our level of creativity and innovation."

Essence of Watering Holes

Watering Holes are not just brainstorming sessions. What has become known as brainstorming has been around for millennia, but it wasn't until 1948 that Alex F. Osborn, a former journalist turned advertising executive, developed the technique and gave it the label *brainstorming.* Adapted from a 400-year-old Indian Hindu process, brainstorming is used in groups to "amass all the ideas spontaneously contributed by its members."[2] It is useful for uncritically collecting as many ideas as possible so that a group can find solutions to a problem.

Now, almost seventy years later, the demand for generating ideas for developing high-value innovations far outstrips the conventional uses of brainstorming. Indeed, some research has shown that conventional brainstorming has limited, or perhaps no, effectiveness.[3] Brainstorming, to be effective, must be structured around answering specific questions, and a great deal of work needs to be done before and after the meeting.[4]

But even this is not enough. What is required is to not only generate new ideas, but also to *exponentially improve the customer value* of potential innovations through the *compounding* of ideas. Without a process to improve your initial ideas—over and over—your creative flash of genius will never result in a new innovation.

Innovations require a synthesis of many ideas to succeed, including the new product or service, enabling technologies or capabilities, barriers to entry from competitors, a compelling business model, and essential partnerships. Ad hoc brainstorming meetings cannot lead to the depth of understanding that innovation demands.

Watering Holes leverage reactions from a disparate audience in a structured format to create customer value. Only by regularly tapping into the genius of this extended team will new, high-value innovations be created rapidly enough to keep up with the exponential economy.

Structured, Recurring Meetings

One of our colleagues called these meetings Watering Holes because they are market-focused, structured, and recurring meetings that give substance to all the "business animals." Although not a direct analogy to watering holes in nature, the name stuck. They are a "safe place" where people can try out new initiatives and gain ideas, feedback, and resources. Participants include the technical, business, financial, and legal staff plus outside experts, as needed. The meetings bring together multidisciplinary teams to get all the perspectives needed for success. Groups interested in a specific market segment, product, or service meet every two to eight weeks to strengthen their value propositions. There are generally five to twenty participants, including:

- a facilitator who announces, sets up, and runs the meetings, and allocates funds

- champions and their innovation teams

- coaches, partners, and customers from inside or outside the enterprise, as appropriate

- Jungle Guides—market domain experts
- Other multidisciplinary contributors, such as technical, business, and legal staff

Watering Holes are not populated with bench sitters—everyone is expected to participate. If you have a value proposition for a new project or business initiative, you are expected to participate regularly to continue refining your ideas. Watering Hole meetings respect people's time. The presentations are crisp and to the point. Each champion of an innovation stands up for the team and gives a five- to twenty-minute presentation of his or her value proposition, proposal, or innovation plan, followed by ten to twenty minutes of specific feedback.

Watering Holes help make the presenter's value proposition compelling and quantitative. You assist the presenter in constructing and improving their Elevator Pitch (Chapter 8) and eventually their proposal or innovation plan (Chapter 9). Watering Holes capture the essence of brainstorming by opening up a free flow of ideas—but always centered on improving the presenter's value proposition to create new customer value.

Profound realizations often come out of Watering Hole meetings. For example, as described earlier, the BBC gained clarity about the need to create greater value for their customers, the public. The BBC is now focused on creating the maximum *public* value.

Metrics

Watering Holes can help you identify *important* customer needs, not just interesting ones. If you have an idea for a project, the first criterion is whether it is important—in terms of either existing customers, market size, impact on the world, or other organizational criteria. Communicate your criteria for important needs to all Watering Hole participants. Otherwise you waste time and disappoint presenters.

In a nonprofit organization working on housing for the homeless, for example, "important" might be judged by the number of people who receive quality shelter. Within a company's financial department, "important" might mean reducing billing cycles by 20 percent. For new commercial products or services, "important" usually relates to market size, growth, and profitability. Whatever the goals of the organization, communicate clear metrics for success. For example, "We are only interested in businesses that can have revenue of more than $50 million per year" or "We want to have a return on investment of more than 15 percent per year."

Watering Hole meetings cost time and money, and the returns must justify the investments. The meetings have to be efficient: everyone in the room should gain from being there. At SRI we form Watering Holes only when we are seeking to create significant new customer value, not for everyday tasks, although we still apply our other concepts, such as NABC value propositions, to all activities.

Watering Holes come and go in response to the marketplace. At SRI some Watering Holes last for six or twelve months and are then disbanded because the opportunities have been fully explored. Others go on for years because we have made a long-term commitment to a major business area.

A Learning Environment

The Watering Hole structure ensures individual and organizational learning. It takes effort and a great deal of experience to understand how to be successful at value creation, and Watering Holes provide a powerful forum for this learning to occur. Consequently, just showing up at a Watering Hole is not enough. As the BBC did in preparing for the Watering Holes in our workshops, there are pre-meeting actions, specific steps during the Watering Hole, and post-meeting actions. All these are designed to maximize learning and position champions and teams for the next improvement in their value propositions.

Before coming into the meeting, the champion and team will have talked to potential customers and iterated their presentation with others. Until you have talked to potential customers, all business discussions are academic. No one at SRI receives significant investments until he or she has interacted with customers and understood their needs.

Presenting a value proposition is hard at first because there is much to discover. For example, essential learning in Watering Hole meetings comes from watching others struggle with and then master a new concept, such as "market positioning" or "business model." When you are in the audience, frustrated because you cannot figure out someone else's presentation, you see how hard it is to be clear and compelling. Since we all struggle to understand the same general issues, Watering Holes help us become more open to improvement. And then, when one of your colleagues figures out a key point, you are better able to appreciate and learn from it. Because Watering Hole discussions can be confusing to presenters at first, it is helpful to have a coach or partner who meets with the champion of the project both before and after the meeting to help improve the concepts presented. The coach also helps by taking notes for the champion during the Watering Hole discussion, when ideas are being thrown out as fast as people can talk. This helps because we have found that the most the presenter can remember at the Watering Hole is about 10 percent.

Watering Holes are meant to enhance the champion's and team's value proposition, not to destroy the champion's confidence. It is inevitably frustrating and sometimes mortifying when at first you are still learning and feeling insecure. Participants build the champion up, not down, by being allies and friends.

Watering Holes provide motivation. The feedback spurs you on and your competitive juices kick in, providing the stimulus to do better. Once someone does a good job on an aspect of the value proposition, the entire group wants to exceed that mark the next time. We have not found a faster means of teaching innovation concepts.

Providing Organizational Support

Watering Holes are an excellent way to eliminate organizational boundaries, which are major barriers to innovation in most companies. You can invite participants from a variety of disciplines and projects whose input may be valuable. Inevitably, someone at the Watering Hole will know someone across the organization who has a bit of critical information. Champions can be empowered at the meetings to go across the company to talk to that person. It is also important to invite *potential* champions who have not been through the process, who can become motivated and inspired by the Watering Hole experience, and who can then begin their own innovation journey.

We provide modest financial resources for use by the facilitator in charge of the Watering Hole meetings. These funds are to send people on trips to visit potential customers and partners, hire market experts—Jungle Guides—build models and simulations, and generally support the early first steps in creating a new innovation. Once a promising innovation has been identified, more substantial financial resources are provided.

SRI uses different names for Watering Hole presentations as they evolve. These names help set expectations. A research team will come to a Watering Hole seeking early input on an initial idea, and gather input. We call those "jam sessions," just like impromptu music jam sessions. In a jam session, an invention might be applied to a potential business opportunity. There was, for example, an early presentation on electro-active polymers, a plastic-like material that expanded when a voltage was applied. It had the potential to replace electric motors, valves, generators, and speakers. This idea eventually turned into a new company, Artificial Muscle, Inc.[5] After the jam session, they did more research and presented it multiple times, finally moving toward a more complete innovation plan.

These later presentations are called "rehearsals," because you are still working hard to make it better: parts don't fit together and other parts are missing. Finally, when you are eventually getting ready to meet with clients, you have a "concert"—a presentation of a complete innovation plan. You are trying to fine-tune aspects of the presentation and make the hard choices about what to include and exclude. In all types of Watering Holes, you receive constructive feedback and continue to improve your value proposition.

Getting Started

Here are some additional rules of thumb to help kick off the Watering Hole process and have it be effective:

- Have a clearly articulated initial goal: find a market segment where your expertise and market needs best align and where the thematic focus has value for all participants.

- Announce the schedule of the meetings and provide a formal agenda: set out the framework for the meetings so your teams can see the trajectory of what you are attempting to accomplish.

- Store and share your achievements: keep either a loose-leaf binder or a website containing all the presentations and backup materials.

- Make market analysis a regular presentation topic: keep building a broader and deeper understanding of the market segment's ecosystem.

- Include prospective partners: create value with your partners to gain their knowledge and buy-in.

Encourage a broad range of participation: important innovations usually happen at the boundaries of different disciplines. Look for those colliding exponentials described in Chapter 2, which open up new opportunities for white-space innovation.

Every time we introduce a new process into SRI we ask, "What is the alternative?" For example, "If we don't focus on delivering the highest customer value in the shortest time, what should we focus on?" This question might seem rhetorical, but sometimes people have blurted out "Me." Asking for the alternative often sharpens the issue to help make better decisions. As you run your Watering Holes, whether they are face-to-face or virtual, you should be asking the same question at each step. But we hope that you will emulate the BBC, institute Watering Holes, and experience the joy of building value propositions that compound in quality to match the changes occurring in the exponential economy.

more ideas for faster value creation:
origins of Linux

"The better we get, the better
we get at getting better."[1]

Douglas Engelbart

A World of Ideas

Can hackers around the world create valuable software in their spare time? Perhaps surprisingly, yes. The operating system Linux, first a curiosity and then a movement, is now a valuable product. Linux has been accepted by IBM, Hewlett-Packard, and Novell as the operating system for their servers. Its development exemplifies how more ideas lead to faster improvements. Rather than a "closed" system, such as the proprietary operating systems of Apple and Microsoft, Linux was initially put on the Internet and developed by programmers throughout the world. Its incremental improvements to the kernel developed by Linus Torvalds in 1991 made Linux more valuable with each iteration. It is now a powerful and stable operating system moving onto desktop computers as well as corporate servers.[2] Linux is a revolutionary development that demonstrates the power of open-source innovation.

Wikipedia.org is another example of innovation that taps into the genius of the world's population to gather ideas. Wikipedia is an open-source encyclopedia available on the Internet that has emerged as a daily tool for those interested in comprehensive, in-depth, encyclopedic knowledge. Wikipedia is available in more than one hundred languages and it is growing exponentially, with an increase in articles of more than 100 percent per year and with active contributors growing at more than 200 percent per year. There are several million articles available, and the average article has been edited more than fifteen times. Some think, "If it is not on Wikipedia, it's nowhere." To see the emerging ascendancy of this approach, consult Wikipedia.org on a topic of your choice and you will find top-notch articles that are both readable and informative. It seems remarkable that such an extensive, high-quality document could be produced with a minimum of supervision and control. But it has a huge advantage that makes the difficulties pale by comparison: It has thousands of people from around the world making contributions. More important, it taps into some of the best minds in every field, who occasionally write articles with quality that matches professional encyclopedia staffs.[3] As the Japanese proverb says, "None of us are as smart as all of us."

In the commercial world, Procter & Gamble is one of the leaders in leveraging innovation ideas from outside. Rather than relying only on their in-house research labs, their "Connect + Develop" program seeks "opportunities to connect with innovators from around the world."[4] As Larry Huston, the VP of innovation and knowledge, says, "Imagine opening up your borders for free trade. P&G employs 7,500 people in its R&D division, but there are 1.5 million scientists throughout the world with expertise in P&G's areas of interest. It doesn't take a genius to figure out that if you can engage the brains of your 7,500 plus the key ones from that 1.5 million, you can build better products."[5] P&G now has more than 35 percent of their innovations coming from around the world, with a goal of 50 percent within a few years.

Another example from Eli Lilly, the pharmaceutical company, is InnoCentive. It is an Internet-based forum of 80,000 technologists in 175 different countries who collaborate for a fee with companies to find innovative solutions to complex problems.[6]

Exponential improvement, as discussed in Chapter 2, is possible when four conditions are in place: (1) an important need is to be satisfied, (2) there are sources of new ideas, (3) a compounding process is in place that builds on previous advances, and (4) appropriate human, financial, and other resources are available to accomplish the work. Rapid exponential improvement requires identifying the best ideas quickly and compounding them often.

Consider again Linux and Wikipedia. First, both address important needs that have no visible upper limit of achievement. Second, both tap into a continuous supply of new ideas from around the world. Third, with Linux, Torvalds provided the first kernel that allowed rapid, compound growth—individuals added improvements to his core kernel and one improvement built on another. In the case of Wikipedia, its software framework and organizational design provide the platform for one idea to build on the next. Fourth, the resources needed to drive these exponential-improvement communities are provided primarily by people of passion, people who want to make contributions to topics they care deeply about. The *Wall Street Journal* described an example where a security vulnerability was discovered in Linux. A person in Germany e-mailed a friend in the United States, who got others involved, and within twenty-four hours the problem was fixed. Tapping into knowledge, ideas, and skills over the Internet can be powerful when people care.[7] Contrast this with your organization when it has a similar problem. How long do you think it would take to get it fixed—days, weeks, months, never?

Virtual Watering Hole

As we have illustrated in the examples given, the Internet provides a way to create virtual Watering Holes. It is also a useful metaphor for the power and difficulties that arise in high-performance teams, as we will discuss in Chapter 11.

How useful can the Internet be at helping us gather new ideas? Most are familiar with the observation that the potential value of a radio or television station is proportional to the number of listeners or viewers. If you add another viewer, the value of the radio or television station goes up by one. This is often called "Sarnoff's Law" after David Sarnoff, the pioneer of radio and television and the first president of the Radio Corporation of America (RCA).[8] Commercial radio and television stations measure their total audience each month to determine how much they can charge for advertisements.

The Internet has much greater potential value and can be used in many more ways. Bob Metcalfe, one of the inventors of Ethernet, noted that in a fully connected network, where everyone can communicate with everyone else, the number of user connections goes up as the square of the users.[9] Since each connection in a network is potentially valuable, if you connect N users you get the potential for N-squared value. The value of such networks can increase rapidly even if you use only a fraction of the total possible connections.

The number, and potential value, of interconnections actually goes up *much* faster than Metcalfe's Law, which we suspect he realized. If you count all the possible interconnections, they actually grow *exponentially,* as pointed out by Reed.[10] Since the increases are exponential, the number of connections *far exceeds* Metcalfe's Law. We refer to this result as the "law of exponential interconnections."

So what's the difference between Metcalfe's Law and the law of exponential interconnections? Metcalfe's Law counts up all the *direct connections* between the users, like those on a telephone network. The law of exponential interconnections counts all those connections, plus all the ways one can *broadcast* information over a network to groups. Thus, the Internet has all the communication advantages of the telephone, radio, and television. In addition, there is the ability to multicast to any selected audience group. All these modes of communication are valuable, and the total number of connections grows quickly as the number of users grows.

In general, we will *not* be able to take full advantage of all the communication possibilities described by the law of exponential interconnections. But even if a user taps into a tiny fraction of the Internet's potential, the law of exponential interconnections demonstrates why this person is using a communication resource of unprecedented value.[11] Imagine if someone could figure out a way to use all this potential. Not surprisingly, many groups are working to do just that.

In the exponential economy, creating high-value innovations requires getting insightful ideas fast. As the discussion above illustrates, the Internet provides a powerful new capability to facilitate the conditions needed for exponential improvement, first as a direct means to access new ideas and second as the basis for entirely new business models, including ways to collect more ideas, as P&G does. Chapter 11 will describe additional models for how you can exploit the potential of the Internet to gather ideas and achieve exponential improvement.

Getting Started

Whether you are in a face-to-face or virtual Watering Hole, you begin the process of creating customer value for a new innovation by rapid iteration with others to get the specific answers needed. Iteration is composed of (1) writing down your initial NABC value proposition, (2) getting feedback on it and collecting more new ideas, (3) synthesizing the ideas collected and revising the value proposition, and (4) continuing this process over and over.

Writing down your value proposition is essential; talking about your ideas is never enough. Our colleague Susan Gauff, who was head of human resources at Sarnoff Corporation and now runs her own company, always pointed out, "If it's not written down, it's not real." She's right. Only after your ideas are written down can you show them to others and collect their ideas for the next iteration of idea gathering and knowledge compounding.

Because change is inevitable, keep it simple at the beginning. Start new initiatives by creating only four PowerPoint slides, one each for need, approach, benefits per costs, and competition. The first version will be poor, but don't worry. By keeping your NABC value proposition short in the beginning, you will also avoid falling in love with your initial approach. What counts is being open to improvement while driving the improvement process. Only after you have repeatedly talked with customers, deeply understand the competition, and have refined your unique approach will you have an excellent value proposition.

Create illustrations, pictures, computer simulations, and prototypes of your solution so that people can quickly grasp and understand your value proposition. In the beginning, you will want to use multiple rough sketches to encourage a broader array of ideas from your prospective customers and partners. When you get closer to the solution, you will make these illustrations and prototypes as accurate as possible in order to get specific feedback. In all cases, be open to new ideas and the prospect of completely discarding your initial proposal.

Present your value proposition to experts, colleagues, and friends, allowing them to help you add to and refine your ideas. Present them in Watering Holes, at lunch to colleagues, to your spouse at night, and to anyone else who will listen. Each person will be interested, knowledgeable, and confused about different things. They will challenge you to answer questions in ways that are simple and convincing. These different perspectives are valuable because, as computer scientist Alan Kay says, "Point of view is worth an extra 80 IQ points."[12] Each of us has a lifetime of experience and knowledge that makes us a "genius" when thinking about certain problems.

As a champion, it is your job to collect these "extra 80 point-of-view IQ points" and synthesize them into a strong value proposition. Each person forces you to present your case more effectively, and each iteration brings more IQ points to the process. We call this tapping into the genius of the team. Try out your value proposition on the most demanding and knowledgeable people you can find, including your significant other. We do. Most people are more than happy to share their knowledge with you. It is flattering to be asked.

No NIH. Encourage your teammates to avoid the Not Invented Here (NIH) disease. In the exponential economy, most of the great ideas are neither yours nor your organization's. To test this, just plug your favorite idea into Google and see how many hits you get. It will inevitably be there. You distinguish yourself by collecting these individual ideas and synthesizing them into a new solution.

Customers Aren't Virtual

Get out of your office; hit the road. The best source of information about whether your value proposition is on the right track is your prospective customers and partners. They don't live in your office and they aren't all on the Web; you need to call them and visit them. Iteration with customers and partners should occur before you attend Watering Hole meetings.

Watch your customers. Rather than relying only on what they say, actually observe them. For instance, if you are working on a new way to present classroom information using handheld computers, observe students in a variety of classes rather than just asking them what works. If you want to design a new word-processing program, sit with people using a product and see where and when their frustration arises. If you want to design a new shopping cart, go to the grocery store and watch shoppers in action.[13] If you want to improve health care delivery, go to a hospital and watch patients; see what their total experience is.

Simple observations can yield vital results. For example, if you were an airport designer, imagine what you could learn just by spending a few hours in an airport. You would see people sprawled on the floor by electrical outlets and roaming around with electrical cords dangling from their hands trying to find a place to plug in their cell phones, portable computers, and other electronic devices. Consider as another simple example your last visit to a hospital. Could you easily figure out where to go? Open your eyes and ears to collect the best ideas.

Philips Medical Systems employs the concepts of customer observation to understand the potential for new customer value. They build sophisticated medical imaging systems, such as magnetic resonance imaging systems (MRIs) big enough to image a full person. One of their managers took the *SRI Discipline of Innovation* workshop and then went back to design a new table that patients would recline on before entering the MRI. As the manager said, before the workshop they would have told an engineer to take out 20 percent of the cost. But they sent a team out to see how the table was used, both by the patients and the doctors and nurses. What they discovered was a family of new ways to improve the benefits provided while reducing costs.

Ask Questions

Most people talk too much and listen too little. As the saying goes, "God gave us two ears and one mouth—use them proportionally."[14] The purpose of visiting customers is to get them to show you and tell you things you don't know. In addition to observing them, ask questions and listen intently for understanding. In critical meetings, assign someone the job of listening to the customer. Talking *and* listening are too hard when it is a tense meeting.

The sooner you get out and interact with customers, the better. They will give you critical information about the marketplace and customer needs. Friendly prospective customers will often volunteer to be one of your "iteration partners," encouraging you to come back time and again as your value proposition becomes more focused, interesting, and valuable. Let your future customers and partners be your free expert consultants.

Yvon Chouinard, the founder of Patagonia, quotes Kristine McDivitt about asking for help after he turned the company over to her to run. She said, "I had no business experience, so I started asking people for free advice. I just called up presidents of banks and said, 'I've been given these companies to run and I've no idea what I'm doing. I think someone should help me.' And they did. If you ask people for help—if you just admit that you don't know something—they will fall all over themselves trying to help."[15] That is our experience too.

Our colleague Len Polizzotto says that when he takes the members of his team in to talk to a customer, he tells them just before the meeting—only partly kidding—"If anyone says anything other than to ask questions, you will be shot." Prospective customers love to talk about what they are interested in and what they most need. Remember: "The person who talks the most has the best time." Make sure your customers and partners have a *great* time.

Polizzotto has an excellent example of the power that comes from asking informed questions. Once, when he was working for a large company, his team was developing a new medical device requiring Federal Drug Administration (FDA) approval. Normally, such approvals take 18 to 24 months. Len didn't want to wait that long. He immersed himself in a crash course on the FDA approval process using outside experts and then had a friend help him set up a meeting with the entire FDA approval team. During the meeting Len *only* asked questions about what each person needed for approval. He was after ways he could make their jobs easier, which would then make his life easier. As each person described what was required, Len took notes on a flip chart to confirm his understanding. Since he had done his homework before the meeting, he was able to ask informed questions as they went along. When everyone had spoken, Len went around the room one last time to double-check his understanding of what it took to obtain FDA approval. He submitted the paperwork for his device; it was approved in only five months.

Success requires deep market knowledge. Obtaining an understanding of a market space—with its business models, players, suppliers, competition, and potential disruptions—takes years of experience. It calls for someone who has walked in the customers' shoes. Polizzotto, for example, consulted with experts about the FDA approval process. If you do not have that experience, you need a Jungle Guide. As our colleague Carmen Catanese once said, "It's a jungle out there; we'd better get a guide." The Jungle Guide is a content or domain expert who leads you through the jungle. You can get lost in a jungle and you can get lost in the marketplace; use a Jungle Guide to make the trip less risky and more fruitful.

It takes work to gather, observe, and synthesize the insights required. When you start this process with prospective customers and partners, it will be, at first, confusing. But a remarkable thing will happen: After a few awkward and perhaps slightly embarrassing meetings, when you really don't have all the answers, you will quickly progress. Your meetings will begin to take on a very different tone. And at this point you will be able to focus your attention on those remaining issues that represent real difficulties and risks. *You* will rapidly become *the* expert.

Ask informed questions. Since almost no one else does, you will distinguish yourself while gaining critical information.

Remove Risk—Be Demanding

The exponential economy is demanding, so you must be demanding of your value propositions and innovation plans. Our colleague Walter Moos says, "If you are standing still, you are moving backwards." True—in today's world you are actually going backward at an exponential rate.

Unfortunately, the majority of project teams in organizations are not nearly demanding enough about developing compelling, quantitative value propositions. The result is billions of dollars of lost resources and untold hours wasted every year. Even in Silicon Valley, the entrepreneurial center of the world, only about one of every five new start-ups has significant success.[16] Many start-ups fail because the individuals involved were not prepared. They have an idea, they write a quick business plan, they raise an insufficient amount of money, they spend it on the wrong things, and they fail. These are hard lessons to learn.

One of your objectives in developing a value proposition is to reduce risk. Successful entrepreneurs and individuals in organizations do their homework. They get solid answers to the critical questions, minimize risk, and thereby maximize value. Consequently, they spend their scarce resources only on those few elements of the plan that are risky. Think of learning how to play a musical instrument. Many musicians fritter away their practice time rehearsing the parts they already know. The people who rapidly improve are those who work on the parts they can't yet play. Risk can take many forms:

- market risk
- technical risk
- people risk
- financial risk
- business-model risk

Don't be afraid to admit what you don't know. Put the issue squarely on the table to focus attention on the need to address it. Over time, you will collect the ideas needed to figure out the answer.

This point is worth emphasizing. Most people spend large amounts of money and time before a clear value proposition has been developed and obvious risk has been removed. *This is always a mistake.* Until a compelling business direction is established, much of the initial effort will be wasted. We believe in moving quickly but with extreme parsimony until a compelling value proposition has been developed. Money should initially be spent *only* to develop the value proposition and the business model, and to remove those elements of the plan that represent risk. Once the direction for the initiative and major areas of risk have been reduced, it is much easier to raise the money needed to complete the project. Moving ahead without a compelling value proposition is the biggest mistake we see companies and inexperienced entrepreneurs consistently make. It is a project, product, and new company killer.

You might be asking yourself, "What is the compelling value proposition for my current situation?" It can be a bit unsettling to ask this question. If you don't know your current value proposition, you can't fully contribute. If you do know it and it is not compelling, then you may either be wasting resources or be vulnerable to the competition. Fix it.

It's a Journey

Your initial value proposition will be incomplete and flawed. The first version of Linux, for example, which Torvalds published in 1991, was usable only by serious hackers. It wasn't until March 1994 that version 1.0 of Linux kernel was released.

The path to success is seldom straight. Being wrong in the beginning doesn't matter. As the design firm IDEO says, "Fail often to succeed early." Rarely do you end up where you thought you would when you started. You shouldn't expect to, since in the beginning you have not talked to, studied, or observed your customers and competition. You don't know which approaches are feasible. And you have not developed a business model that is viable. Successful innovations require a journey. You need to listen carefully to the marketplace and you need to be adaptable. Your original vision might have been to end up, metaphorically, in Chicago, but the right destination for your innovation might be San Francisco.

Early failures cannot be avoided; don't think that you will be the exception that gets an easy ride. Look at the notebooks of people who have accomplished great things and you will see failure after failure. Pasteur, for example, generated one strange theory after another. But he kept learning, rejecting theories that did not work, compounding his knowledge, and eventually he succeeded in inventing vaccination and the process of pasteurization.[17]

Find a partner, create your first four PowerPoint NABC slides, talk to some potential customers, present at a Watering Hole, and continue the iteration process afterward. The journey will be a lot more fun and productive than you ever imagined.

Exponential Improvement

At the beginning, improvements to your value proposition will hardly be noticeable. But when all the pieces start coming together, you will experience an exponential-like improvement in your value proposition— which you need to match the challenges of the exponential economy. As you continue, your research and the ideas you get from others will compound. The "curve of exponential improvement," as indicated in Figure 7.1, shows how the quality of a typical value proposition develops over time.

Note that you don't iterate just once or twice, you do it many times, improving at each stage. Most people don't iterate enough or fast enough. They stop after three or four iterations. Nor do they seek out all possible feedback from customers, partners, and others. They get discouraged, protect themselves from input, and give up. Your value proposition is the first step in creating your new innovation. Once you get the answers needed, you go on to create a new innovation plan, raise the resources needed, develop a prototype product or service, and then deliver the final product or service to your customers in the marketplace.

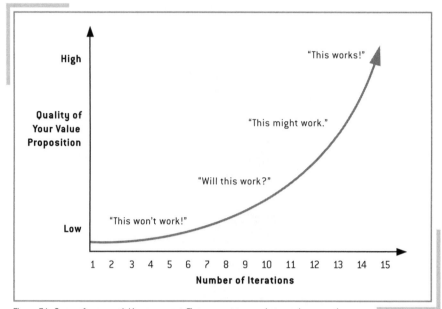

Figure 7.1: Curve of exponential improvement. The comments are what people presenting value propositions often say as they make the journey of value creation. When you have answered the fundamental questions and improvement slows down, it is time to move to the next stage in developing your innovation.

Edison continuously improved his value propositions—often hundreds of times. When a colleague of his observed that he had again failed to find a filament that would not burn out in a lightbulb, he said, "I didn't fail, I just found something else that didn't work."[18] But eventually he succeeded. Reach out to your colleagues and iterate, iterate, iterate until you create a compelling value proposition. If you do, you too will experience the curve of exponential improvement, uncover more customer value, and increase the likelihood of a successful innovation, just as Linux did.

YOUR eLeVaTOR PITCH: HOW HDtv BeGaN

Getting to Yes

Could it be done? For two decades Japan led the development of
HDTV. Every HDTV "first" was owned by NHK, the national broadcasting agency of Japan. Among their accomplishments was the first
practical HDTV system, which they called MUSE.[2] It was a superb accomplishment, but it was an analog system that used a great deal of
bandwidth. The question every video laboratory around the world was
trying to answer was whether an all-digital system was practical. It
needed to have full HDTV resolution, good motion rendition, and use
roughly half the bandwidth of MUSE. This wasn't an easy question to
answer in the early 1980s. It's another example of trying, as Edison did so
well, to figure out whether all the critical parts can be made to come together at the right time. If you are too early you can waste hundreds of
millions of dollars; if you are too late you can miss a multibillion-dollar
opportunity.

Our team had worked on HDTV for years, first as part of RCA Laboratories and then as part of Sarnoff, SRI's subsidiary. We had been developing persuasive arguments for why digital HDTV was now possible.

Our customer at the time was Thomson Electronics, which had bought RCA from GE. We had a chance to make our case during a major review at Thomson's digital video laboratory in Germany with the senior VP for research, Eric Geiger. We were given only a few minutes to present our case for starting an HDTV program in the United States. Geiger is a remarkably perceptive and courageous manager, but many of the people at the meeting were unreceptive to the idea of a U.S. digital standard. Consequently we wanted to avoid getting into a protracted debate about the minutiae of HDTV, which could have derailed our proposal.

We decided to keep our presentation extremely short, and to present it as an Elevator Pitch composed of six questions, which we would address directly to Geiger. If he said yes to all six, the result would be the start of the program. With a splendid colleague, Glenn Reitmeier, we worked for months on the questions and possible answers. The goal was to make it easy for Geiger to say yes. Here is a truncated version of our elevator pitch to him with the six ingredients in bold:

THE HOOK: "Do you believe that consumer electronics will be revolutionized by digital technology over the next ten years?" Geiger said, "Yes."

NEED: "Digital HDTV, with five times the number of pixels and high-quality digital sound, would be an exciting new consumer product that would create a 35mm film experience in the home and replace every TV. For example, just the total home TV market will be well in excess of $20 billion per year. Do you agree?" Geiger said, "Yes." "Excellent."

APPROACH: "We believe by 2003 that we can practically compress and transmit at consumer prices a 1.2-billion-bits-per-second high-definition signal over today's United States 4.2 MHz analog television channels, which can be allocated by the FCC. Our approach would use an advanced version of MPEG-2, which our simulations show can provide the performance we need. Are we on the right path?" Geiger said, "Yes."

BENEFITS PER COSTS: "Establishing the U.S. HDTV standard with your current 25 percent market share represents a five-billion-dollar-per-year new business for your company with two to three times the ROI of today's products. It will provide the foundation for multiple waves of new products and services, such as camcorders, CDs, and consumer games, each of which will result in several billion dollars in new revenue. Do I have this correct?" Geiger said, "Yes."

COMPETITION/ALTERNATIVES: "We have demonstrated a number of capabilities that distinguish our system from the competition, including 20 percent or better image quality and new digital services such as interactive advertising. We could start this program in Europe, but that approach would not have credibility with the FCC, which must approve the system. The United States is the beachhead market that must be won first. Reasonable?" Geiger said, "Yes."

THE CLOSE: Then do you agree that we should immediately start a U.S.-based HDTV program?" Geiger said, "Yes."

We asked our questions, and to all six Geiger said, "Yes." Then we sat down before everyone in the room fully realized what had just happened. Clearly, we left the technical details out of the above summary, and you might feel that the justification for HDTV is obvious now, but it wasn't at the time. The hard part of this presentation was figuring out how to get back to yes if he said no. Fortunately, because we had done our homework, that was not necessary.

Getting Noticed

Ideas have always been a dime a dozen, but never more so than today in the exponential economy. We are all bombarded with "the new new thing." Getting the attention needed for success requires that you distinguish your ideas by their clarity and value. If you are the champion of a new innovation, one of your jobs is to raise the financial and human resources needed to get your project completed. This always means that you must convince someone—a company president, a board of directors, a venture capitalist, or a government program manager—that you have a good idea. In most situations, the number of possible projects far outstrips the available financial resources. SRI board member

Henry Kressel, who is a managing partner at Warburg Pincus, the world's largest venture investment firm, says, "If there is a quarter on the table, there will be fifty people reaching for it." Venture capitalists, for example, receive hundreds of proposals for every few they fund. There are a lot of hands out for their "quarters."

Distinguish yourself by giving clear, succinct presentations. In a nonprofit agency, if you can succinctly state the great social value of your proposal, others can join with you. In the business world, if you can tell your story in a few minutes, you have a much better chance of being funded.

The Elevator Pitch

An Elevator Pitch is a pithy summary of your value proposition that can be told in one to two minutes. It grabs the attention of a potential customer, partner, employee, or supervisor who can't forget it and wants to learn more.

One of the hardest things to teach people—especially bright, accomplished people—is the importance of succinct and vivid communication. They know all the details and they feel that leaving any out will destroy the integrity of their presentation. "Do you want me to mislead them?" That's categorically *not* what we have in mind. What we are after is finding the core message and presenting the best possible value proposition. We want to determine "What *must* our audience remember about our message?"

Keep your message as short as possible. Remember Strunk and White's succinct admonition from *The Elements of Style*: "Omit needless words."[3] In the movie *Amadeus,* the emperor said to Mozart, "Your composition has too many notes!"[4] That was ironic, because Mozart's music is unique in that it has *no* extra notes; every note matters. If you take one out—even one—you feel the loss. Think of Mozart as your role model. Take out all those extra "notes" until you can give an Elevator Pitch that would make Mozart proud.

If your audience doesn't understand, it's your fault. It is too easy to say, "They should understand," or "I need more time to explain it."

These are all cop-outs. As a champion who accepts responsibility for the success of your program, you must be willing to iterate and develop your value proposition until it is compelling to all on your team. In the exponential economy, if you cannot communicate with zip and force, there won't be anyone listening. Sequoia Capital—a leading Silicon Valley venture-capital company that helped form Apple, Google, and Oracle—says, "One test we often apply to a new business is the ease with which it can be explained. If someone is able to summarize his company's plan on the back of a business card, it usually means that he will be able to describe its purpose to employees, customers, and shareholders. It suggests they truly understand their business."[5]

Melanie Griffith starred as Tess, a hardworking secretary in the movie *Working Girl.*[6] She wants to make it in the world of high finance, but she is stuck in her role. Her boss, Katherine, often accepts good ideas from her without giving credit. One day, by accident, Tess discovers that Katherine is planning to steal another of her ideas and present it to Mr. Trask of Trask Industries, Katherine's boss. This idea would open up a new market in radio, allowing Trask to improve their market position without competition. A few days later Tess, Katherine, and Mr. Trask are all in the lobby of the office building. Tess has a confrontation with Katherine, Mr. Trask steps into an elevator, and just as the elevator door closes and Mr. Trask disappears from sight, Tess says, "Did she tell you about the hole in the account?"

"The hole?" Mr. Trask is hooked. He opens the elevator door and steps out. Mr. Trask and Tess go into an empty elevator and Tess is told, "You're on."

She has just over a minute to sell Mr. Trask on her idea during their elevator ride. She is succinct and shows Mr. Trask some of her research and how "the hole" could be closed to Trask's advantage. When they get off the elevator and Katherine rejoins them, Mr. Trask asks Katherine where she got "her idea" and she can't answer. She is fired and Tess wins the day on the strength of her Elevator Pitch. Hollywood understands the importance of a good Elevator Pitch.

Your crisp elevator pitch must exude value. What do we mean by that? The goal should always be to present opportunities whose value is so obvious that the audience *must* take advantage of them. When you have a compelling presentation, you prove that you are focused on the fundamental issue of creating value in the marketplace. You also prove that you will be able to communicate that vision, not only to investors, but also to customers and employees. If employees do not clearly understand the vision of the company and the value being created, then the chances for success are limited.

What does an Elevator Pitch look like? Fred Fritz was president of a disposable hearing aid company called Songbird Medical Systems.[7] Here is Fred's Elevator Pitch to a prospective user:

> Are you having trouble hearing? You're not alone.
>
> Tens of millions of people suffer from hearing loss and find the hassle of getting an expensive hearing aid too much trouble.
>
> For the first time, disposable hearing aids are available.
>
> We offer a disposable hearing aid that costs one dollar a day. It has the best-quality digital sound and, because you throw the units away every month, it is made of a soft material that fits comfortably, securely, and invisibly in your ear. You can buy them over the counter in your drugstore.
>
> By comparison, high-quality hearing aids cost several thousand dollars today, and they have to be fitted by a doctor. Because they cost so much, they have to be made out of hard, durable materials that do not fit comfortably or tightly in the ear, which also makes the sound quality suffer.
>
> Would you like to see one of our disposable hearing aids?

Keep it brief. Every team should be required to prepare an Elevator Pitch for its proposal.

The Elevator Pitch Template

The template for an Elevator Pitch has three parts: a hook, a core, and a close. It has a hook to get interest, a core composed of your quantitative value proposition to tell your story, and a close to ask for action to move to the next step. An elevator pitch looks like:

- **Hook:** to get their interest
- **Core:** your NABC value proposition
- **Close:** action to get to the next step

Your Elevator Pitch must be repeatable by others. Executives, venture capitalists, and leaders in government and universities cannot be expected to remember long, complicated arguments. If they cannot walk out of the room after meeting with you and state simply and clearly why their company or organization *must* move ahead with your project, you have failed. Your project will get lost in the morass of normal business. Good venture-capitalist firms, for example, look at several thousand ideas a year and invest in about a dozen. Stand above the rest; start with a compelling, value-laden Elevator Pitch. And make it brief.

We all know when someone is *not* compelling—the vision does not resonate, you cannot see your role in the activity, and you are not convinced by the value proposition. Here are some ways to develop and improve your Elevator Pitch.

Have a Hook

Advertising people know the importance of having a "hook" that connects your idea to something that is important to the individual, as when you say,

- "Are you having trouble hearing? You're not alone."
- "Each year, one hundred thousand people die because of adverse drug effects."[8]
- "More than two thousand ideas have been patented for new mousetraps and only two are really used."[9]

People relate to stories, metaphors, and humor. Tell a vivid story; it will stick in their minds. When we talk about how Douglas Engelbart and his small team created the foundations of the modern PC user interface, including the computer mouse, people remember. His story crystallizes the idea of innovation teams making an impact, and maybe, as a result of the story, you will want to know how you can create an innovation team too.

Failure stories also resonate. When we describe how, even in Silicon Valley, more than 80 percent of start-ups fail to produce a significant return, it sparks interest. How might that be improved? Hooks inextricably link the listener with the value proposition and activate their interest or curiosity. Provide the audience with a hook.

The Core—NABC

In Chapter 5, "It's As Simple As NABC: How Liz Got Her Big Job," we gave you the essentials of the core of your Elevator Pitch. The NABC value proposition focuses you on the audience need—not your own need—and on the approach you propose, the benefits per costs you can provide to your audience, and an examination of the competition. The NABC is your quantitative value proposition tucked in the middle of your Elevator Pitch.

Be Quantitative

One of the most difficult things to do, but one of the most persuasive with an audience, is to be quantitative. Rather than saying, "How would you like to improve your workshop offering to the organization?" you say, "Would you be interested in a 20 percent increase in enrollment over the next six months?" Because being quantitative is so rare, it helps you to stand out from the crowd. Further, it forces you to become clear and specific about your value proposition; after all, you don't want to make a promise you can't fulfill.

"Faster," "cheaper," and "better" are not quantitative. Faster than what? How much cheaper? Better than what? And are the amounts significant? You can have a computer chip that is faster than another, but does it matter? For example, 0.1 percent faster probably doesn't matter; it might take 100 percent to sell successfully. You need to demonstrate that your improvement will create *significant* new value for the customer.

Imagine a college basketball scout giving a report to his coach. He says, "I have found a terrific new center, named Kareem, for our team. He is tall, he scores well, and he's a nice guy." Compare that with "Coach, I have found a new center named Kareem for our team. He is 7 feet, 2 inches tall, he has scored 43 points per game, he has a 3.7 grade point average, and he is president of his class." Specificity is powerful.

When talking about market size, give numbers. When talking about product specifications, list them. When talking about profitability, say how much. People like numbers. Think of baseball and football statistics.

If you don't yet know the specifics, tell your audience you don't know and take a SWAG.[10] Put down your best estimate. Say, "We are still working to get the data, but we expect to be 20 to 50 percent less expensive." Over time, you can improve your estimate. But you must at least give ballpark estimates so your audience will have some idea of what you are talking about. Don't compound ignorance with ambiguity.

People in human resources and in nonprofit work find this even more challenging than others. Something feels "wrong" about providing quantitative details. Yet when you do, you are forced to be absolutely clear about what you are offering, and it will stick better in the listener's mind.

Close: What Are You Asking For?

All too often, the close for beginning presenters is to lower the voice, break eye contact with the listener, and then awkwardly sit down—relieved to be done. Slinking away is not an effective close. A good close must accomplish something. After all, why are you talking to *them*? What are you asking for?

- another meeting
- funding
- additional partners and employees
- a reference to another person

Ask yourself before you begin, "What do I really want here?" Is it just to survive the one-minute pitch or to accomplish something? If you don't want a specific outcome, don't waste anyone's time.

Practice and Make It Contagious

Both the message and the delivery must be practiced and improved by suggestions from others. You don't want to give a canned sales pitch. Refine, recast, redo—iterate, iterate, iterate. When you communicate and it doesn't resonate with the team members, or outsiders, refine your pitch. Talk to others about how to cast the message the next time. *None of us gets it right the first time,* so expect to change and refine as you go. Here is a test: Can your friend, who doesn't understand your business or job, understand what you are saying?

Passion sells. Former secretary of state General Colin Powell said, "Enthusiasm is a force multiplier." Winning presentations are given with passion. Passion is a visible sign of your commitment. It is tangible proof that you are a champion: someone who can be believed and followed. We do not mean that you must be flamboyant. On the contrary, presentations can be understated and said softly. But others can tell if you are committed and enthusiastic about spreading the gospel of your project. And passion is contagious—when you make the ideas important and relevant, they almost spread by themselves. Others think, "Gee, if this is so important to them, maybe I could be part of this too."

The nice thing about contagion is that it cycles back to you. The next thing you know, one of your team members is in a meeting giving a passionate plea for your project. The contagion reverberates throughout the team and the organization.

As you begin working on your first Elevator Pitch, Figure 8.1 will remind you of the necessary components. Fill in the blanks and begin the iteration process.

The exponential economy moves fast. It is based on a flood of information and ideas; it is noisy. To distinguish your project, develop a compelling Elevator Pitch and present it succinctly and with conviction. Your Elevator Pitch is your NABC value proposition with an opening hook and action steps in closing. The Elevator Pitch works best if you have a gripping hook, so others can connect with it. Through practice, you can give a confident, compelling presentation with passion. When all these ingredients come together, your enthusiasm will spread throughout your innovation team and beyond. People invest in and join champions who speak with clarity and passion about important projects.

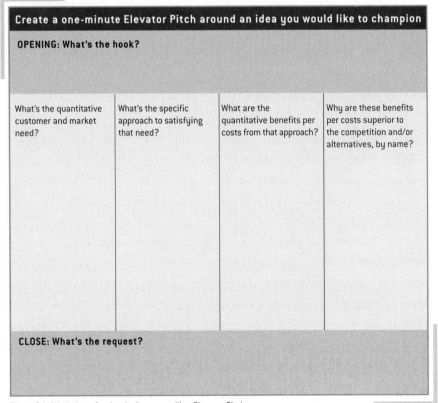

Figure 8.1: Worksheet for developing compelling Elevator Pitches.

CHAPTER 9

YOUR INNOVATION PLAN: *from the ski slope to the firehouse*

"MAKE SURE YOUR INNOVATION IS A PAIN KILLER
AND NOT JUST A VITAMIN."

David Ladd, venture partner at Mayfield Fund

A Lifesaver

When the first responders—fire, police, and other emergency personnel—entered the World Trade Center on 9/11, cellular phone service and electricity were down. Tragically, they were unable to communicate either within the towers or to comrades on the outside. The need for such emergency communication is increasingly being met by Packet-Hop, a spin-off company using an innovation pioneered at SRI. In a simulation of a terrorist attack on the Golden Gate Bridge, PacketHop was able to seamlessly link thirteen different agencies to provide the kind of communication that was missing at the World Trade Center.[1]

The journey of PacketHop began with a ski slope and ended with first responders, and illustrates central features of all innovation plans. It started with SRI's notion that technology might be developed to circumvent wireless phone companies. This new technology would link wireless cell phones directly to one another, bypassing the phone company's cell towers. The telephone message would "hop" from one phone to another until it got to its destination, eliminating infrastructure cost and allowing "always on" communication.

During early Elevator Pitch presentations at SRI Watering Holes, it looked like real-time direct communication between mobile devices would be technically possible. It would take, of course, much more investment to bring this into reality. On the basis of this first assessment of the opportunity, the initial PacketHop team moved to the next step: specifying the first market. The larger market, of replacing telephone company wireless traffic, would have to wait, because that market would only emerge as large communities of users had these phones. We had to start to "cross the chasm" from early adopters to mainstream consumers.[2]

Michael Howse was hired as an entrepreneur-in-residence (EIR) from U.S. Venture Partners to study the marketplace and examine possible uses for the technology. What was needed was a "beachhead" market— one where the technology could be used successfully and then leveraged to open up other new, larger markets. For each possible beachhead market, the need was assessed, the approach to meet that need was detailed, the benefits per costs spelled out, and the alternatives and competition examined. The types of markets studied varied from use in corporate networks, to wireless carriers extending into their "dead zones," to games and entertainment (such as students all playing an online game together on campus at the same time), and to sensors in bridges and highways that communicate highway conditions. At one point, ski slopes were considered. If you are a skier, you know how difficult it is to find your partner when you go down different runs at varying speeds. The idea was to rent PacketHop phones at the ticket counter, which would allow skiers to talk and send text messages and videos to others without anyone else hearing.

But in one of our Watering Hole meetings, it was decided that this was just a vitamin—something nice to have but not a "painkiller," not a migraine eradicator. In fact, if you talk to skiers, you find that many already use nonprivate channels on two-way radios. But more important, conventional cell service was fast expanding to more ski areas, decreasing the need for PacketHop.

What kind of need would qualify as a painkiller? After lots of discussion, Watering Holes, and more research, the needs of first responders became apparent. As firefighters, emergency medical technicians (EMTs), and police enter a disaster scene, the inability to communicate with one another can enormously increase their risk and lessen their effectiveness. Thus, the firemen entering an emergency situation, such as the World Trade Center towers, with PacketHop phones would form their own network, passing calls from one fireman's phone to another and to those outside the emergency location. These are capabilities that cannot be addressed with two-way radios. Clearly, the inability to communicate was pain that needed to be alleviated; PacketHop now had a beachhead. From identifying that need, to creating the first live demonstrations, to signing the first customers, PacketHop is becoming a valuable innovation.

Why Have a Plan?

Why, after all, do you need an innovation plan? Simply because if you are going to invest time and money, you want to be successful. According to the U.S. Small Business Administration, a good plan will improve your chance for success by 50 percent when forming a new business.[3] Whether you are developing a novel health care project for your nonprofit agency, thinking of a new approach to freshman education at a university, or rolling out a new product, you want to *tell a complete story with specific benchmarks to measure your progress.* You describe the opportunity you see and how you will use resources to tackle it. It doesn't matter whether it is to start a new project, a new venture, or a new government program—a plan is required. For the government, it can be a document that says how you will address the requirements listed in a request for proposal (RFP).

Outsiders to Silicon Valley hear about the $8 to $12 billon[4] spent there each year on new ventures and assume that it is a place full of risk takers. But venture capitalists do not throw money at every fascinating idea that comes their way. On the contrary, good VCs work incredibly hard to *reduce* risk. Good VCs, like good entrepreneurs, are *risk reducers,* not risk takers. When you have a comprehensive business or project plan, you are reducing risk. After all, if you are going to spend all that time and money, don't you want to win?

When you first hear "innovation plan," it probably conjures up the image of a 300-page document loaded with charts and Excel spreadsheets. As a result, many people freeze in their tracks. But in Silicon Valley a typical innovation plan looking for millions of dollars will be ten to fifty PowerPoint slides. The writer Hal Plotkin tells this amusing story: "One of my editors at *Inc.* magazine had a brilliant idea: Call up top Silicon Valley venture capitalists and ask them for a list of the ten best business plans they'd ever seen. So I called the usual suspects: Kleiner Perkins' John Doerr, Yahoo! backer Mike Moritz at Sequoia Capital, Cisco Systems' first major investor, Don Valentine, and legendary San Francisco semiconductor company financier Art Rock, among others. I popped the question. And none of the financial A-team members could think of a single written business plan, let alone ten, that had made an impression on them. Some of them even laughed at the question. 'They're usually a waste of paper,' one told me. 'You're kidding, right?' asked another."[5]

We agree. A good innovation plan is based on quality, not quantity. It's not a fifteen-pound stack of paper. It is a few PowerPoint slides that are the result of an incredible number of iterations and hard work.

If you are like most people seeking money, you think you already have your approach figured out. The plan is just padding to justify your approach. This almost always fails, and then you have to back up and make huge changes to make it realistic. It is more efficient to consider those key elements needed for success *before* you start the innovation, project, or new business.

We know of one new business supplying insulated panels for homes where the financial projections were basically created out of thin air, showing a profit on shipping alone of $120,000 the first year. Yet after nine months it was apparent that not enough homework had been done, and in fact a loss of $15,000 had been created. The projections didn't factor in license fees, insurance, workmen's compensation, and a host of other hidden fees trucking companies pay. A lot of time has been wasted trying to redefine the business, all in retrospect, whereas a good innovation plan, with more research, would have uncovered these errors up front.

An innovation plan is an expanded version of your Elevator Pitch. Thus, you already know the skeletal elements of an innovation plan. The hard work is not in producing more pages but in adding critical detail to make it complete. The plan is a document, typically using PowerPoint slides for investors, or a text document for banks. Of course, each prospective investor has different needs. You can be certain, however, that few investors will read a 300-page document, even if they say they will.

Fundamentals of Innovation Plans

Your quantitative value proposition is the heart of the innovation plan, along with additional information, including risk-mitigation. Below is an outline of the central elements of an innovation plan your investor or organization will want to hear. Depending on whether you are forming a new project, a new product or service, or a new venture, some of the ingredients will change. Your organization may also use a different format and order,[6] but the fundamentals are:

- Opening
  - *Vision and mission statements*[7]
  - *Business objective (short enough to fit on a business card)*
- Need
  - *Overall market space (size, players, business models, disruptions)*
  - *The specific market segment (customers, size, growth)*
  - *The important need (where the "pain" is)*
- Approach
  - *Product or service description (the "painkiller")*
  - *Product or service development and production plan*
  - *"Golden nugget" that allows a differentiated and superior product or service (technology, human business, partnerships, etc.)*
  - *Business model (how you will make money)*

- *Product or service positioning (how the product will be sold and to whom)*
- *Financial plan (investments, revenues, profits, and timetable)*
- *Staffing plan*
- *Risk mitigation plan*
- Benefits
  - *Customer benefits per costs, including Value Factor Analysis*
  - *Investor benefits (including profit and return on investment)*
  - *Employee benefits (profit sharing, equity)*
- Competition or alternatives
  - *Competition now and in the future (by name)*
  - *Competitive advantages and barriers to entry [IP, partnerships, etc.]*
- Next steps

You are familiar with most of the elements listed, but a few need additional explanations.

Overall Market Space

The starting point for every new innovation, whether it is incremental or transformational, is to understand the overall "market space." In your market space, you will find partners, competitors, vendors, distributors, and a host of other players. You will find different business models by which organizations make money. Your space may contain specific government work rules and regulations, such as those promulgated by the FDA for therapeutic drugs. In addition, you should be aware of technological, political, and organized labor developments that might jeopardize your initiative.

Within your market space, there is a smaller "market segment" that contains a "beachhead" where you can start, just as PacketHop did.[8] It is for these customers that you define your first product or service. The beachhead is often referred to as the "white space." This is where you have identified an unmet need, you can address that need, and there is minimal immediate competition.

PacketHop had to consider the market space for mobile communication, which includes conventional phone companies, mobile communication companies, Internet service providers, software application organizations, and many portable device companies providing cell phones, PDAs, and application-specific devices. One market segment within the mobile space was for new PDAs that do not require a conventional communications infrastructure to operate. Each device would transmit the signals from one PDA to another, which is what PacketHop enables.

The beachhead market that PacketHop selected addressed the needs of first responders. The product was a PDA-like device with a specific software application. For first responders, this product is not merely something nice to have (a vitamin). Rather, for first responders, the solution is a must-have (not just a painkiller but rather a lifesaver). Applications beyond the beachhead market will eventually be developed.

The interconnected relationships existing in any market space[9] define an "ecosystem," since, just as in a jungle, there are many interconnecting relationships between participants that allow the ecosystem to thrive. These relationships are initially hard to understand and may even be invisible to outsiders. That is one reason we recommend getting someone early on who understands your ecosystem—a Jungle Guide.

Innovations disrupt their ecosystems, like the introduction of a new predator in the jungle. Transformational innovations can either rearrange or even destroy the existing ecosystem. Wal-Mart, for example, by saving the average U.S. working family more than $2,000 per year, has created enormous customer value. But it is also having a major impact on mom-and-pop retailers.[10] The bigger the innovation is, the bigger the disruption. Don't be surprised if, as you modify your ecosystem, the other players in it fight back with a vengeance. Right now, for example, Google is a new large predator attempting to rearrange the desktop-computing ecosystem, where Microsoft has been the 800-pound gorilla for the past several decades. Like most 800-pound gorillas, Microsoft doesn't like being disturbed.

A trap many organizations fall into is "preparing to fight the last war" rather than charting the ongoing changes in the ecosystem. With or without you, the ecosystem changes. If your innovation plan only uses current realities without forecasting the future, you may find yourself like the associates of David Sarnoff who, in response to his request to invest in radio, said, "The wireless music box has no imaginable commercial value. Who would pay for a message sent to nobody in particular?" Or H. M. Warner, producer of silent movies, who said in 1924, "Who the h___ wants to hear an actor talk?" Today we hear newspaper CEOs saying, "Who the heck wants to get their news from the blogosphere?"[11]

"Golden Nugget"

Every new innovation must have a competitive advantage, ideally a *sustainable* competitive advantage that will last for many years. This advantage might come from a new, enabling technology, a relationship, a novel manufacturing process, or a new business model. At SRI we call this key to success a "golden nugget." While it might still need refining, it allows you to create new, compelling customer value. And as in panning for gold, you are always looking for something extremely valuable that others don't have and your customers want.

For example, one "golden nugget" behind Google is its search engine, which directs search queries based on the number of Web users who link to a site. The more popular the site, the more likely it will show up in a Google search. This idea makes rapid Internet search possible. In the technology world, it also helps if you own patents or other rights so that your golden nugget creates a barrier to a competitor's entry. Patents can be an important part of a golden nugget for an innovation.

Golden nuggets come in many forms, including new business models. One of the most famous was pioneered by King Gillette. Before 1900, razor blades needed continuous sharpening. The incessant sharpening meant shavers were inconvenient for customers to use. Gillette realized that great customer value could be provided by selling long-lasting "safety razors" with inexpensive disposable razor blades. The razor could be sold at a loss, but over time the blades would make a great profit. This was a business model golden nugget. His innovation also included the development of a method for manufacturing the disposable blades out of thin sheet steel, a second technical golden nugget. Gillette's business concept is now called the "razor and blade" business model. It has become a classic that has been widely copied. Gillette provided his customers with a better shave, increased convenience, and lowered costs.

Worcester Polytechnic Institute's (WPI) renowned engineering school "sends more undergraduate engineers and scientists abroad than any other American university."[12] The students get real-world experience in their field and in another culture, and as a result of this golden nugget, WPI ranks first nationally in the percentage of science and engineering graduates with relevant international experience. If you were going to hire an engineer fresh out of college to be successful in the global exponential economy, this exceptional educational nugget would catch your eye.

In Silicon Valley a golden nugget can be an enabling technology that opens up new applications, products, or services. "Artificial muscle," for example, is a plastic material that expands and contracts when a voltage is applied. It can be used in robots, speakers, valves, and many other applications. For example, as a replacement for small conventional motors, it can operate at a tenth of the weight and a fraction of the cost. It is a technology that addresses existing multibillion-dollar markets with profoundly better solutions while opening up large new markets. It is a true golden nugget, the basis for a new company, Artificial Muscle, Inc.[13]

Our colleague Norman Winarsky once proposed a golden nugget scale. As with gold panning, the size of nuggets can vary and be evaluated as:

INCREMENTAL INNOVATIONS: No golden nuggets

- State-of-the-art products or services with a unique feature
- Examples: Toothbrushes with non-slip grips, flip chart paper with handles, a magazine on trendy home furnishings, mobile phones with a simple address locator

NOTABLE INNOVATIONS: $1/2$ golden nugget

- A distinguishable advance in either design, technology, process, or business model
- Examples: RAZR cell phone, disposable hearing aids, the first quality ice cream, a unique restaurant franchise

SIGNIFICANT INNOVATIONS: 1 golden nugget

- An important advance in some combination of either design, technology, process, or business model
- Examples: iTunes with iPod, electro-active polymer motors, an effective cancer drug, Southwest Airlines

TRANSFORMATIONAL INNOVATIONS: 2 golden nuggets

- Two important advances in some combination of either design, technology, process, or business model
- Examples: Original Gillette safety razor, Wal-Mart, Google, Toyota production system, lightbulb and electrical distribution system, Microsoft Windows

In these examples, the potential impact increases greatly from top to bottom. It has been suggested that innovations are better thought about on a logarithmic, Richter-like scale.[14] Incremental innovations might have an impact of 1, notable innovations 10, significant innovations 100, and transformational innovations 1000.

These examples are, of course, illustrative and you may put them in a somewhat different order depending on how you evaluate their merits.

Consider, for example, a mobile phone with a simple address locator, it may be better than anything on the market, but it is still not optimal because others will quickly imitate it. It is an incremental innovation—the kind that must be repeatedly made with established products to stay in business. On the other hand, the iPod created both a new way of downloading music through iTunes and a way of playing it on a terrifically designed product—two good innovations that created a solid golden nugget. But Apple still has to keep on running fast with one new innovation after another to keep ahead of the pack. Artificial Muscle's patented electro-active polymer, which has the potential to replace conventional motors in many consumer products, is an excellent golden nugget based on a revolutionary new technology. It has the potential to rearrange the motor and actuator industry. Finally, again consider Gillette, who hit the bull's-eye because he had both a revolutionary new technology (thin blades) and a revolutionary new business model (razor and blades). His two golden nuggets rearranged the entire shaving industry and in the process created a multibillion-dollar company, which was bought by Procter & Gamble in 2005.

There are many opportunities for innovation: new customer products, customer experiences, customer services, supply chain, business structures, administration systems, and organizational structures.[15] As indicated above, innovations are best and most sustainable when several key ideas constructively converge, as in a new product with a new business model. Peter Drucker was quite clear about this: "[A] characteristic of knowledge-based innovations . . . is that they are almost never based on one factor but on the convergence of several kinds of knowledge, not all of them scientific or technological."[16]

Product or Service Positioning

Every new product or service must be positioned within its ecosystem, explicitly communicating its specific benefits to consumers in its market segment. For example, advertising for Corvettes positions them to appeal to males who want a flashy, high-performance automobile at a moderate price. Second, positioning also refers to how the product will be sold. Will Corvettes be sold directly by GM or through distributors? Successful positioning allows the organization and its partners the opportunity to develop an effective business model and to succeed financially.

Business Model

Your business model describes how your product, service, or venture will make money for your business, your partners, and your investors. Sometimes the business model is straightforward. If you wanted to go into the quality ice cream business, you might open a store and sell ice cream directly to customers who walk in. That is how Häagen-Dazs was started in the Bronx in 1961 by Reuben Mattus, who chose a name that conveyed the image of a quality European ice cream. Soon Mattus wanted to expand by selling ice cream through grocery stores. This is harder, because grocery stores generally have existing relationships with other ice cream manufacturers. Someone in the ice cream ecosystem needs to be displaced by removing their products from the grocery store shelves. Along the way, you must find a way to manufacture sufficient quantities of ice cream, get it to the grocery stores, share your profit with the distributors and stores, and still make a profit for your company. Bumping someone out of the ecosystem costs time and money. Until you understand the ice cream ecosystem you don't know how hard that will be and whether it will cost too much for you to run a successful business. Häagen-Dazs figured this out; it is now in thousands of stores in fifty-four countries.[17]

We learned the hard way how challenging it can be to develop an acceptable business model when SRI's subsidiary, Sarnoff, was trying to create digital cinema. We got excited about this opportunity because it is easy to show that digital cinema can have much higher quality than that of today's 35mm projected film. To prove this, we did side-by-side comparisons at Warner Bros. for industry leaders, where the best 35mm film was shown side by side with projected digital video. These demonstrations, to the amazement of those in the industry, showed that the quality improvements from digital were striking.[18] Visually it was a slam dunk. This was a good start, but had we identified a vitamin or a true painkiller? Unfortunately, this was just a vitamin, because the current 35mm film quality was, from an industry perspective, good enough.

The pain for the movie industry was not in picture quality but rather in the cost of distribution and the huge losses resulting from illegal copies being sold around the world. When 35mm film copies are distributed to thousands of movie theaters, it is easy for pirates to steal one and then flood the underground market with cheap knockoffs. The loss to the industry is billions of dollars a year.[19]

Now we were getting somewhere. This was a real pain point. In addition, we had the golden nugget technology that would allow secure distribution of digital movie copies by satellite, saving hundreds of millions of dollars each year. When hundreds of millions of dollars are up for grabs, there is generally a way to make the opportunity attractive to all parties involved. But not in this case. It turned out that there was no viable business model. The interlocking contractual relationships between the studios (which make the films), the distributors (who ship the films to theaters around the world), and the exhibiters (your local theater that makes money selling popcorn) could not be changed. The digital cinema opportunity will have to wait until the future, when either the need becomes great enough or these contracts expire so the ecosystem can be disrupted.

To develop your business model, you have to do your homework. Until, as in the case of electronic cinema, there is a viable business solution, there is no opportunity. You must understand your ecosystem and the relationships between the parties. This also illustrates the point that we repeat again: be parsimonious about spending human and financial resources until you can see your way to a potentially viable business opportunity. Getting some of the major risks out of the way quickly is just smart. In the case of electronic cinema, we had access to all the product and service solutions, but we couldn't make the business model work. Fortunately, we didn't spend much money to learn this hard lesson. But electronic cinema will happen. HDTV in the home, with better than 35mm-cinema quality, and declining theater receipts will eventually force the movie industry to provide a dramatically superior visual experience.

Tips for Success

Value Factor Analysis

While your projects or products can have "value," it is only when you establish the superiority of your value relative to the competition that you make headway. People always have alternatives; there is always competition. It may be another company, another project, another organization, or just the alternative of doing nothing.

In the Appendix you will find a detailed treatment of Value Factor Analysis, which, as described in Chapter 4, allows you to numerically compare the relative customer value of your product with that of others. It helps to determine whether you are only marginally better or have a product that will make an important impact. If you are serious about your business plan, consult that Appendix and use Value Factor Analysis.

Picture, Mock-Up Simulation, Prototype

When you move from the Elevator Pitch to a full innovation plan, your customers will need more of a feel for your new product or service. Give your customers a picture or mock-up of what you are creating, even if the mock-up doesn't work. In today's world of computer-aided design, there is no excuse for not having a professionally rendered picture or mock-up of your new product.

Of course, if you have a complete working prototype, all the better. For the electronic cinema demonstrations described above, we lashed together parts of a high-definition system we had developed earlier. It did not have the full quality of the solution we were proposing, but it was good enough to prove that electronic cinema would be dramatically better than today's 35mm projected film.

In our workshops, we always insist that participants create mock-ups of their business proposals using whatever is available, whether it is cardboard, Erector Sets, or Tinkertoys. For example, when the BBC participants in our workshops produced a family of cardboard models of new digital devices that would transform the broadcasting business, and presented them to one another in a Watering Hole, they "got it." Similarly, members of ITRI, the prominent research institute in Taiwan, came up with an interesting new mobile device. Their demo came complete with blinking lights. Everyone could feel this new invention working, although it was just a mock-up. Mock-ups work for your audience, even when they don't actually work!

When developing new consumer products or services it can be particularly difficult to identify unmet customer needs. These needs are often vaguely understood and cannot be crisply articulated by customers. This enduring problem is becoming even more challenging as we progressively satisfy tangible needs and focus more on intangible ones, such as styling, experience, and identity. In these cases having a trial product or service—a mock-up, prototype, or simulation— becomes especially important. The first, quickly developed, version of the new product or service can be based on the intuition and experience of your innovation team, which should include professional designers. This trial product or service then becomes a value-creation tool to allow customers to better articulate their hidden needs. As they use it, play with it, share it, and compare it with alternatives, they can begin to communicate what works and what is missing. In effect, your customers are co-inventing the new product or service with you. Google, for example, is doing this by putting beta versions of new products on the Internet to get feedback from users. Your trial product becomes a critical part of your quantitative value proposition and, eventually, your innovation plan, because it facilitates rapid improvement through the compounding of new ideas.

Innovation Plan Presentation

A compelling innovation plan can be delivered in twenty minutes or less. During our workshops, we require that everyone answer all the questions listed previously in five minutes. The answers given for each point can only be a few sentences long. Nevertheless, these are always some of the best presentations our participants have ever heard. Making it short forces you to identify the most important points, which are usually all your audience wants to hear at first.

Learn as much as you can about what is valuable to your audience and then modify your pitch appropriately. Relate to them. We have all been in the audience when a presenter talked at length only about what they were interested in, while showing no concern for the audience. Did you think highly of, or want to partner with, that presenter?

People are grateful when you either e-mail or call in advance to find out how you can best meet their needs. This increases your chance of being effective and shows them that you are respectful of others. Some questions to ask yourself about your audience are:

- What does your audience want to achieve?
  - *What are their needs?*
  - *What keeps them up at night?*
  - *How can you help them?*
- What are their backgrounds?
  - *How much do they already know?*
  - *What will they want?*
- What do you want them to do?
  - *How can they help you?*
  - *How do you want them to react or follow up?*

Beyond the Innovation Plan

Once you get your innovation plan developed and funded, there is still a long journey before a new product or service is successfully delivered into the marketplace. Nevertheless, throughout the product-development process, your value proposition continues to be a critical tool as new opportunities for customer value creation become available as you learn more. We used it, for example, with leaders from Philips Medical Systems, discussing all the elements of value that go into a new MRI machine costing over a million dollars. Many of the features that create the most value for doctors and their assistants have to do with the convenience of using the system. Many of these value-adding service and experience features are understood only as the system is designed, prototyped, and first used. The opportunities for innovation never stop.

It always comes back to the customer. As with PacketHop, as you move from the ski slope to the firehouse, your innovative product or service should be "brought to market in an innovative way, supported in an innovative way, branded in an innovative way, but in the end [it has to be] . . . an approach that builds enduring relationships between the company and its customers."[20]

DISCIPLINE
3

INNOVATION CHAMPIONS

a CHAMPION:
the mayor of kellyville

NO CHAMPION, NO PROJECT, NO EXCEPTION.

A Journey

John Kelly, a world-famous atmospheric scientist, is also the unofficial mayor of Kellyville, Greenland, a town that, on a busy day, has a population of 10. It is named after him because Kelly and his team have run an important atmospheric laboratory there for years. Through his work at Kellyville, he developed a concept for a new innovation with the potential to revolutionize atmospheric research. The journey to realize this new system helps explain what every innovation must have: a champion. Previously, in Chapters 5 to 9, we discussed the tools to use in value creation. These tools, however, are ineffective without someone actively using them to push the innovation forward. There must be a *champion* who proactively identifies with the customer and who addresses the funding, bureaucratic, political, human, and technical challenges that every innovation faces.

During a recent conversation, Kelly told us why many of the early warning signs of environmental problems can be first seen in the atmosphere. He then described how sunspots impact the upper atmosphere and how those changes can impact the environment and radio communication around the world. He went on to describe issues in global warming and ozone holes, and how little the world understands about these phenomena and the resulting consequences for our future. He became particularly excited about how beautiful the night sky is, especially when you are in Greenland looking at an aurora borealis when the entire sky seems to be on fire.

Kelly is slender with a gray beard. He answers questions carefully, and although he is calm, he exudes excitement about the importance of his work and the deep regard he has for his colleagues. You immediately sense that you can trust him and that it would be a wonderful experience to work with him.

Atmospheric research is difficult because it is hard to measure subtle changes occurring miles above the ground. Information can be obtained using airplanes and balloons, but it is limited. Kelly's interest is in radar systems. Most atmospheric radar systems look like those big radio dishes you occasionally see from the highway or in *Scientific American.* They work by transmitting a radio signal to a point in the sky, which is modified by the atmosphere and then reflected back to the same radar dish. Over many years, atmospheric scientists have learned how to figure out what these reflected signals mean. This has proven to be an effective way to understand the atmosphere, but it has a big problem—the atmosphere moves. Those conventional dishes can move to change the point in the sky where they are looking, but much too slowly to follow the atmosphere because they weigh too much. For a long time, creating a radar system that could move fast enough seemed impossible. But Kelly had an idea to solve this problem and transform atmospheric research. It would take more than fifteen years of hard work before it came to fruition.

The idea behind this revolutionary innovation was to build a radar system where the *signal* moved across the sky while the *radar* stayed still—just the opposite of conventional systems. This new radar would be made from 10,000 little radars several feet on a side and all grouped together in an array as big as two football fields. By timing the signals to each of the 10,000 little radars properly, the signal beamed into the atmosphere could move extremely quickly, and could scan out a two-dimensional video image of the sky. It could easily track the fastest atmospheric phenomena and for the first time collect data essential for understanding our environment.

Kelly and his SRI colleagues had this idea for a revolutionary radar system back in 1988. He assembled his team and they put together their first sketch of what it would look like and do. Even though there was initial excitement about the idea, it was going to be an expensive project, costing more than $30 million to build something as big as two football fields. His customers for this new radar would be fellow atmospheric scientists at the National Science Foundation (NSF) and around the world who are working to understand atmospheric effects on the environment. Kelly had many meetings with them to understand their needs and to learn how to design a system that would collect the most valuable data.

In 1989, Kelly put together a detailed proposal and submitted it to NSF. It was not prepared to fund such a project, but it asked Kelly's team to submit another proposal for a study to prove that the concept could work. Kelly submitted the proposal, the team won the study, and they showed that the concept was feasible. Then there was a long hiatus. Kelly and the team kept on working hard, but the project always missed being funded. Finally, in 1996, a solicitation was issued by NSF, and Kelly submitted a detailed proposal in 1997. The proposal won. Unfortunately, for political reasons, Congress intervened and removed funding for the project. Then, after another long delay, NSF asked Kelly to completely redesign the radar system. To anyone else this would have been a fatal blow, but in retrospect Kelly acknowledges that this review in 1998 greatly improved the system. But again, in response to Congress and just before funding the latest proposal, NSF restructured the funding process. This caused still another delay. Working with his partners he came up with yet another proposal to satisfy both NSF and Congress. Finally, in 2003, the system was funded and part of the system has now been built. It works superbly.

John Kelly is a champion in every sense of the word. He selected an important need, built a great team, and iterated his value proposition hundreds of times. He worked through incredible obstacles and persevered for fifteen years to make his innovation come to fruition.

When we look back, it is hard to believe this project could have succeeded had it not been for Kelly. Fortunately, most innovations don't take fifteen years to come to fruition. But some do. Whether it is an incremental innovation that can happen in months or a transformational one like Kelly's, there must always be a champion.

Kelly's experience is not that unusual. The process of innovation inevitably involves many ups and downs. Even famous innovators like Steve Jobs, CEO of Apple Computer, who now seems to move effortlessly from one innovative project to another, go through many ups and downs before they succeed. In 2005, he gave the commencement address at Stanford University.[1] He talked about how dropping out of college was one of the most important decisions in his life, how he got fired from Apple and why that helped him grow, and what it was like to think he was going to die from cancer. Reflecting on his current success, he said:

> I'm pretty sure none of this would have happened if I hadn't been fired from Apple. It was awful-tasting medicine but I guess the patient needed it. Sometimes life's going to hit you in the head with a brick. Don't lose faith. I'm convinced that the only thing that kept me going was that I loved what I did. You've got to find what you love, and that is as true for work as it is for your lovers. Your work is going to fill a large part of your life, and the only way to be truly satisfied is to do what you believe is great work, and the only way to do great work is to love what you do. If you haven't found it yet, keep looking, and don't settle. As with all matters of the heart, you'll know when you find it, and like any great relationship it just gets better and better as the years roll on. So keep looking. Don't settle.

Every project requires a champion: someone with the skills and commitment to make the project a success. Champions are innovators who master the Five Disciplines needed to rapidly create new customer value in the exponential economy.

What Is a Champion, Anyway?

Innovation champions can come from any part of the enterprise: management, professionals like Kelly, or someone just starting a career.[2]

Champions are builders who are passionate and committed. They stay focused on a vision and inspire their team and partners to work together. They persevere by taking full responsibility. When road bumps occur, as they did in Kelly's case, they come back at the problem in new ways. These characteristics may seem impossible at first, but in our experience almost 100 percent of the people who make the commitment to be champions have significant success. Champions follow their passion. They have an urgent "Let's do it" spirit—as they always fully involve others.

Champions and the teams they work with achieve success by following the value-creation process we have described thus far, starting with identifying an important customer need and driving the value-creation process. Champions continue involving their team, improving their value proposition, and meeting unforeseen obstacles. They continually talk to potential customers and fold new information into the project's value proposition as the project moves toward fruition. The champion in essence has binocular vision—she must simultaneously face the obstacles directly in front of her and keep focused on the long-term vision.

Earlier we mentioned DARPA, the Defense Advanced Research Projects Agency, which is a primary U.S. funding agency for advanced technologies. It hires superb people and gives them the authority, responsibility, and resources to run their programs. There is little infrastructure and almost no bureaucracy. DARPA is a small group of "champions" with a *big* budget. It is not an accident that DARPA was responsible for the first step in building the Internet. J. C. R. Licklider of MIT first proposed a global network of computers in 1960, which he called an "Intergalactic Network."[3] Everyone on the globe would be interconnected and be able to quickly access data and programs at all other sites. In 1962, he became the first head of the computer research program at DARPA in order to champion the Arpanet, which would turn into the Internet.

Unfortunately, many people feel that they work for an enterprise where their best ideas for helping the enterprise will be neither valued nor rewarded. Or they may feel stymied, as Frank Guarnieri was in Chapter 1, when the enterprise will support them but they simply don't know how to proceed.

Companies that welcome ideas not for the value they create, but because they come from someone with clout or a title, leave enormous money on the table. Potentially great ideas go unnoticed because they originate from some "unimportant" person. New ideas from outsiders have no way of getting accepted since they are "not invented here." In the exponential economy, this is dangerous, since as Bill Joy, former chief scientist of Sun Microsystems, points out, "No matter how many smart people you have, most of the smartest people are never in your own company."[4] If employees don't believe they can be a champion like Kelly, then neither they nor the enterprise will reach its potential.

Champions Need Partners

Champions who succeed almost always have at least one other person who helps them launch their innovation. Steve Jobs, for example, had Steve Wozniak.[5] And Frank Guarnieri? He was on the verge of resigning, but as a result of working with us, he improved his value proposition for developing drugs until it led to the formation of Locus Discovery. Without a partner, most of us lose our enthusiasm, limit our vision, succumb to the obstacles, and stop trying. While the champion helps others succeed, she also needs support to step up and model champion behaviors.

While Hewlett-Packard Company started in a garage, neither of the two primary innovators, Bill Hewlett and Dave Packard, was the "lone genius in the garage." Rather, they were both champions for their joint inventions, from testing equipment to their first computer.[6] Their collective genius, along with their team's, launched a company known for innovation around the world.

All inspirational people have others close by to guide, support, teach, and nudge them. One of the things you discover when you have a partner is that you not only produce better results, you have more fun. You of course need to pick the right partner. If you pick someone who is a cynic, for example, he or she will drag you down. What you need is someone with a high level of competence who adds to the activity. After all, you aren't just trying to fill a seat; you are looking for a function to be served. Partners are crucial to all champions, new businesses, and even private life.

Organizationally Responsible

Champions are organizationally responsible. They abide by the values, team norms, and internal structures of their enterprises. They do not run amok. Their performance is reviewed and advised by others, and they are connected and aligned with the overall mission of their enterprise. They are not free agents or mavericks subverting the purpose of the enterprise; they are sources of energy leading it forward.

Being a champion does not mean that you can drop all other responsibilities or recruit people from other activities without approval. On the contrary, champions gather support and organizational responsibility by the power of their ideas, the trust they generate, and the ability to make their enthusiasm contagious. Here are a few more tips:

- *Listen and learn:* All input is positive when the goal is increased value.

- *Fail often to succeed early:* Test your ideas early and often.

- *Ask for ideas before asking for resources:* Keep costs down and interest up.

- *Surround yourself with enthusiastic volunteers:* Recruit for passion, curiosity, and values as well as talent.

- *Build business and financial models early, but be skeptical:* Be quantitative and start with a SWAG.[7]

- *Thank the thinkers, praise the participants:* Share the credit, acknowledge contributions.

- *Trust the process:* Visit Watering Holes and iterate, iterate, iterate.

Some people are natural champions, but most of us need to work hard at developing the attributes and skills listed above. But whatever skills you start with, it takes experience and practice to gain the deep knowledge needed to be an effective champion.

Different Champion Roles

Overall Project Champion[8]

Every successful project and venture is driven by at least one overall champion. Someone must be the force that gets the project going—and keeps it going, from concept, to implementation, to completion. A critical, unique role of the overall project champion is to continuously synthesize new ideas from the extended team. The project champion is, in effect, the overall systems architect who keeps the innovation on track.

Many joint ventures between companies fail for one simple reason: there is not one single person with overall responsibility and ability to conceptualize the entire project. A notable exception was the New United Motor Manufacturing, Inc. (NUMMI) car manufacturing plant, a joint venture between GM and Toyota in Fremont, California. This was successful, in large part, because Toyota supplied the overall project champion with the necessary expertise in quality improvement. Someone is always needed to guide the overall effort.

Champions are the critical catalysts and facilitators of innovation teams. They do not have to personally do all the jobs represented by the Five Disciplines, but they make sure they all get done. We believe that having a champion for each project is a defining criterion in all enterprises. If an employee asks for resources, the first question should be "Are you the champion?" *No project should be started without a champion, because the chances for success without one are almost zero.* No champion, no project, no exception. If the project has to proceed and a champion is missing, one must be found.

CURTIS R. CARLSON AND WILLIAM W. WILMOT

Everyone Is a Champion for His or Her Part

Each person on a team should be a champion for her or his part. In large organizations, for example, each person on the executive team must be a champion for their part, for the overall enterprise, and in individual roles in support of teammates. When a new initiative is being developed, the champion for finance, for instance, will say, "I will immediately take responsibility for all the budget and financial considerations to support your project." Or Paula, the person from Human Resources, says, "There are federal guidelines about equity and fairness rules that we must follow; I'll e-mail you all a summary this evening and call you tomorrow. I will make sure we address these obstacles so we can move forward." The champion for each part automatically steps up, makes offers, and begins to help the team make sound decisions.

Champions can emerge from any level of the enterprise. Because the core responsibilities and skills of champions do not depend on specific professional knowledge, champions can emerge from any position in the enterprise, from the president, to the mid-level manager, to the stockroom administrator. Rank does not make you a champion. You cannot be told to be a champion. Only you can decide to be a champion. If you decide you cannot be the champion, step aside so one can be found.

A colleague at the Sarnoff Corporation, Jeremy Pollack, is one of our role models for how to be a champion. Pollack is a research assistant, five organizational levels down from the CEO. But you would never know that. He is smart, persuasive, and engaging. When Sarnoff has a really tough problem that must be solved, Pollack asks for it, takes complete responsibility for it, and gets it done. When we were developing the U.S. HDTV standard, we would go to the consumer electronics show every year in Las Vegas. Setting up experimental HDTV equipment the size of three refrigerators in a building where union rules prohibit you from plugging in a piece of equipment without a union member holding the plug is a challenge. Pollack was never daunted, even when he had to work all night in an empty exhibit hall to get the equipment to work. It always did.

An enterprise full of champions exudes energy, optimism, and excitement. You can feel the difference in someone saying, "I'm the champion who helps teams stay on track," compared with "I'm in human resources." Each new champion contributes to a positive compounding spiral of possibilities where people succeed.

Champion Rewards

Champions accrue many rewards. They get to make an impact in their careers because they focus on important needs and create value for their customers, teams, enterprise, and themselves. On the personal level, they get to learn new skills and enhance their own professional development. In the exponential economy the rare individual who has mastered the Five Disciplines of Innovation will have the opportunity to be successful in many business settings. In addition, they will likely be employable for as long as they want to work.

As important, champions experience the joy of working and sharing with others, since their passion attracts other talented people who want to join them in producing extraordinary results. Being a champion is simply more fun.

As a champion, when you look back on your career, one of the things you will likely treasure most will be your terrific colleagues, remembering the excitement and camaraderie generated by working on important projects together. In addition, there are great rewards that come from helping others reach their potential. When one champion we know was asked what he liked best about his job, he said, after a thoughtful pause, "I like helping others achieve their dreams."

Goethe had it right: "Boldness has genius, power, and magic in it."[9] Become a champion; become it now.

DISCIPLINE

4

INNOVATION teams

ǥenius of teams:

douglas engelbart and the birth of the personal computer

"NEVER DOUBT THAT A SMALL GROUP OF THOUGHTFUL,
COMMITTED PEOPLE CAN CHANGE THE WORLD:
INDEED IT'S THE ONLY THING THAT EVER HAS."[1]

Anthropologist Margaret Mead

Standing Ovation

If anyone is to qualify as a transformational visionary, it would be Douglas Engelbart, who, more than anyone else, was responsible for the innovation we know as the personal computer. Yet Engelbart's great accomplishments were the result of teamwork, as is the case with most world-changing innovations.

Scientists don't get standing ovations after they give presentations. But on December 9, 1968, at the San Francisco Fall Computer Conference, in front of a thousand researchers and developers, Engelbart and his thirteen team members from SRI did.[2] And for good reason: they described the basic computer functions that are on your desktop computer today.

It was what many believe to be the greatest single demonstration of new computer science ideas ever. During the presentation Engelbart unveiled, among other contributions, the computer mouse, multiple windows, on-screen editing, hypermedia, context-sensitive help, distributed collaboration, and shared-screen video teleconferencing. The demonstration was given at a time when everyone was still using punch cards to interact with computers. On behalf of his team, Engelbart presented a startlingly new and dramatically more human way for people to interact with computers.

He did not just talk about these ideas at the conference; he *demonstrated how they worked*. These concepts were so compelling and so far beyond what the world thought possible that when Engelbart finished the presentation, the thousand people in the audience leapt to their feet and cheered. It was like a rock concert—the *only* major technical talk we are aware of where the presenters received a standing ovation. When technical folks listen to their peers present ideas, they often think, "Well, that's interesting, but I think I can do better." Not this time, though.

Engelbart's presentation was an epiphany. It changed the careers of thousands of the world's leading computer scientists and led to the modern computer interface. These ideas ultimately spawned the creation of the Apple Macintosh, which popularized this revolutionary human-computer paradigm. Now every personal computer uses these ideas.

Engelbart and his small team created a computer-science tour de force that has never been surpassed—it was the best ever. His team was the most productive in the history of computing, which doesn't happen by accident. Engelbart's team used processes that are the archetype for capturing team genius, which trumps the genius of the individual when creating innovations. Had Engelbart's team members been working in isolation from one another, they never would have made such strides.

They did it by first selecting an *important* problem. Their objective was to "make the world a better place by augmenting and extending human intellect." Their project had the potential to make a huge impact by addressing a fundamental human need for greater productivity tools. Even though they were doing basic computer research, they were also working to satisfy the needs of prospective computer customers. This was not a study in abstract computer science.

When you first talk with Engelbart, the thing that strikes you is how gentle and respectful he is. But it becomes quickly apparent that he is a passionate *champion* for his ideas. You are also left with the impression that you have just talked with one of the most determined and committed people on the planet. Asking Engelbart whether he is totally committed to his projects would be an insult. He and his team were enthusiastic, unrelenting advocates for their work and its utility—true champions. He, and they, still are.

The team members worked in a small, common research space and often labored through the night. They may have fought fiercely about the project's ideas, but they respected one another and shared basic human values.

They used a value-creation process, which was one of the team's groundbreaking contributions. They demonstrated that you can greatly improve value by *iterating and compounding ideas.* They took new ideas from all around the world and brought them into the project. All good ideas were welcome. The team believed it was not enough to just talk about their ideas: *ideas had to be implemented and continuously improved.* Because they implemented and shared all ideas, the useful ones were visible to the team, and this inspired them to create even better ideas. Engelbart was fortunate to have had a brilliant team that could implement these ideas, led by Bill English with other key software staff, such as Jeff Rulifson. The hardware behind Engelbart's memorable presentation in San Francisco was itself a singular achievement, for which English and his colleagues deserve enormous credit.

Each day the entire team would see the overall progress on the project and then add new ideas, constantly leveraging and extending the achievements that had come before. Yesterday's product became today's tool to make still another product, as shown in Figure 2.4. They could see which ideas were powerful, and they could develop new insights about how to work with computers to augment human intelligence. The tools extended the abilities of their users, and the users extended the capabilities of the computers. Engelbart called this "bootstrapping." This process was repeated over and over. Each iteration brought more value and created the process that leads to exponential improvement. They were a high-performance innovation team.

Currently Google has embraced some of Engelbart's insights by leveraging the Internet to connect to customers and gather additional ideas. They publish unfinished beta versions of new products so that customers can use them and provide feedback. Google and their users co-invent the final product, which makes them an Engelbart bootstrapping community, where the capabilities of both the products and the users are extended and improved.

Normally, people would hear the Engelbart story and conclude that he was a genius. True enough, but that is not the whole story. Engelbart and his team were successful because of the processes they used. That is the overarching message of and an inspiration for this book. These processes can enable others—both genius and nongenius—to achieve remarkable results too.

Engelbart's revolutionary achievements have been recognized. He has won many of the highest awards in computer science, including the IEEE John von Neumann Medal, the A. M. Turing Award, the Ben Franklin Award, and the National Medal of Technology, America's highest technical honor.

Everyone loves to be given recognition and to be rewarded for their efforts, but during a discussion with Engelbart he reminded us what it was like being at the center of the project. It was hard, frustrating, and often disappointing. Enormous effort and perseverance were required. The achievements did not unfold in a simple, logical way, and there were huge obstacles, both technical and human, to be overcome. Even today, close to forty years later, there are still major parts of Engelbart's vision—such as the power of Networked Improvement Communities to capture a team's collective intelligence to rapidly solve important problems—waiting to be widely used. We will describe these shortly.

The history of these events could have been more satisfying for Engelbart and his team if Discipline 5, organizational alignment, had been in place. For organizations to realize the full potential of their innovation teams, receive the financial returns, and garner the worldwide recognition of their contributions, they must be committed and prepared to take innovations into the marketplace. Engelbart and his team were creating a new industry, but at that time the mission of SRI was not to fully commercialize inventions, so it licensed the computer mouse to Apple, Xerox, and others. SRI learned many lessons from Engelbart. It is now an organization dedicated to the full innovation life cycle.

Goal: Exponential Improvement

The rapid, exponential development of new customer value is a requirement for staying ahead of the competition in the exponential economy. As we have described, exponential improvement is possible when there are four requirements in place: we are addressing an important need, we have access to new ideas, a compounding value-creation process is in place, and the human, financial, and other resources needed are available. You can see that Engelbart's team satisfied these four requirements. By adhering to these principles they created the foundations for hundreds of billions of dollars of new customer value.

In the following sections we will describe in more detail why the collective intelligence of a team can perform at the genius level and how to manage the interconnections within a team for collaboration. Then, we will overview a formal structure called a Networked Improvement Community (NIC), which exemplifies how to tap into the genius of an ever wider community. Finally we will provide a checklist for you to use to evaluate your teams.

Collective Intelligence

Why can productive teams be so powerful? Engelbart referred to a team's "collective intelligence." One of his main objectives was to find effective ways to utilize this potential to get new ideas that would allow his team to create at the genius level.

The military has a clever method of demonstrating the power of teams. It first asks a group of soldiers to solve a problem individually. After the tests are scored, the group is asked to solve a similar problem, but this time *by working together.* The soldiers then compare the results, with the inevitable conclusion that the team does dramatically better than the best individual in the group. This demonstration is simple, but it is extremely persuasive.

Creating new technologies, projects, and businesses requires new visions, new business models, and unconventional solutions. Different perspectives can provide powerful insights about what is possible. The genius of teams emanates from the particular viewpoints and skills we each have—our perspectives. When each of us looks at a situation we transform it into a framework that makes sense to us based on our experience, training, and knowledge. Some of us may think of analogies, others may think of new images, and still others may think of new statements of possible solutions. Our different perspectives can transform a problem that seems impossible to one person into one that is readably solvable.

In mathematics, the power that comes from transforming problems from one mental representation into another is a well-known principle. At least two mathematical representations are usually needed to solve hard problems. One representation allows certain parts of the problem to be solved, while a second representation allows different parts to be solved.

Sometimes it takes many representations to solve extremely hard problems. One such problem was Fermat's Last Theorem,[3] which Fermat proposed more than three hundred years ago. Until Andrew Wiles proved it in 1994, it was the most famous unsolved mathematical problem in the world. Practically every famous mathematician and some famous physicists, such as Einstein, had attempted to prove that it was correct. They all failed. Wiles proved it by making many mathematical transformations. Each transformation solved a part of the problem until the entire theorem was proven. This is what properly constructed teams help us do too.

High-performance teams also help us get out of ruts. When we are working on hard problems, we all get stuck. We end up in intellectual culs-de-sac we cannot easily get out of on our own. Our colleagues and friends help us get unstuck through their unique perspectives. Even Andrew Wiles had a few trusted colleagues who made critical contributions at the end, when all seemed lost.

As in the case of Engelbart's team, where he insisted that all ideas had to be implemented, it is critical to have your ideas written down so you can create an iterative, compounding process. The tool to aid that process is your NABC value proposition. Using this tool makes it possible for others to see where you are, help you get unstuck more efficiently, and add new ideas.

Another positive example of innovation teams comes from IDEO, an award-winning design firm in Palo Alto, California. The subject of an ABC *Nightline* special called "The Deep Dive," IDEO has been responsible for the product design of hundreds of products, such as Palm handheld computers, Panasonic Toughbook computers, and the lifelike whale in the movie *Free Willy*. IDEO even teamed with Kaiser Permanente to improve health care delivery. Putting themselves in the roles of patients, IDEO researchers discovered the need for comfortable waiting rooms, lobbies with clear instructions on where to go, and larger examination rooms with space for three or more people and curtains for privacy.

Rather than rest on the genius of individual designers, IDEO convenes high-performance, cross-functional teams. Each team consists of people from disparate disciplines—such as engineering, graphic design, psychology, anthropology, and medicine—who concentrate on the end users' use of a product. Each team follows a disciplined yet free-form discussion leading to successive stages of development.[4] In the Deep Dive video, IDEO teams:

- focus on the quality of the idea—not the hierarchy of roles
- are multidisciplinary
- explore real life and observe customers in action
- emphasize prototyping
- use a group facilitator to set time frames and boundaries

IDEO is an excellent role model for product design, Engelbart for world-changing research and product development, and W. Edwards Deming and Taiichi Ohno for quality.

High-performance teams tap into the genius that resides in the collective knowledge, intelligence, and transforming perspectives of others. This ability to transform problems into new representations so as to look at them in novel ways is particularly important in the exponential economy, when changes are occurring rapidly in diverse areas of business, technology, health, communications, society, and ethics. These are complex problems. It is increasingly true that major contributions can be made only by teams that cross conventional disciplinary boundaries. Champions must assemble teams that are just large enough and contain the skills needed to solve these problems.

Implementing Virtual Watering Holes

Innovation teams leverage the power of the law of exponential interconnections both in face-to-face and virtual settings to get access to the best ideas. Linux, Wikipedia, and Procter & Gamble, as we mentioned in Chapter 7, use approaches to tap into the world's online population. Each is like a huge virtual Watering Hole to collect knowledge, guided by a few principles for participating.

Engelbart described a model that takes advantage of this networked potential, which he called a Networked Improvement Community, or NIC. A NIC is designed to capture a community's collective intelligence in order to solve important problems rapidly. It provides another model for either face-to-face or virtual Watering Holes where the ideas in this book can be aggressively applied.

A NIC attacks an important problem by simultaneously working on different but overlapping parts of the problem. It is assumed that all participants in a NIC are fully interconnected by a network, either human or electronic, like the Internet. The results from the overlapping parts are continuously fed as input to all other participants working on the problem.

Consider a NIC where the goal is to improve K–12 mathematics education. This is a community because everyone has a common interest, namely, helping children learn. The people involved are passionate about using their specific skills. They provide the resources needed to make the NIC function. The NIC has three families of overlapping activities providing new ideas and input into all the others. For K–12 education, the first level might be teachers who are educating children in mathematics. They would use results from the NIC and report back both what works and requirements for unmet needs that must still be addressed. The second level of the NIC might represent work on tools that can improve the teaching task, such as computer simulations of mathematical concepts. A third level might be basic educational research, such as new insights into cognitive learning.

First-level users provide feedback and needs to the second-level tool builders, who then provide feedback and needs to the third-level basic researchers, and vice versa. The goal is to create an iterative, knowledge-compounding process among all members of the community to achieve rapid, continuous improvement. Engelbart points out that a NIC must also have a *knowledge repository* to store information about what people have done so that others can build on it (e.g., like Wikipedia or a notebook). It also needs an *interpreter,* who is a person or system whose job it is to find and organize information within the knowledge repository and across the networked community (e.g., like Google or a facilitator).

Obviously, there can be other NICs feeding into our education NIC to add even more new ideas. For example, there can be additional NICs for government policy, disabled children, and home schooling. There can also be meta-NICs for improving the process of innovation. As we have discussed, this form of continuous collaboration can lead to exponential-like improvements. Eventually, there will be computer software agents that will help us find other partners and information to further speed up the idea-collecting process.[5]

These simple ideas are very powerful. SRI International conducts an education research program called Tapped In[6] using this general structure with excellent results. It has a community of more than 16,000 educators. Teams make an impact in ways that would be impossible through more conventional approaches. We asked Judy Fusco, the Tapped In facilitator, for some examples of how teachers were helped to be more effective through participation in Tapped In. She mentioned B. J. Berquist, a teacher at the Loysville Youth Development Center, a male juvenile correctional facility in central Pennsylvania, who listed a suite of teaching skills and classroom content she had gained from Tapped In, such as ways to share Internet content with her students.

None of the examples Berquist gave was huge, but Fusco concluded, "As Berquist and I talked, we realized that although the help given sometimes seems small and inconsequential, it all adds up to have a large effect." This is, of course, a virtue of compounding based on continuous improvement.

But running a NIC is not easy. Fusco points out that all the human issues in any normal organization are magnified in proportion to the accelerated rate of improvement. There must be someone who acts as not only an interpreter but also a mediator to solve the human and other practical issues that continuously arise.

Fusco added, "What we have learned from our experiences is that although a community can do a great deal on their own, a community does need a formal, paid leader for day-to-day management and operations. There needs to be someone, or a team of people, to make decisions and enforce them by 'locking acceptable pages to prevent unwanted changes.' The paid community staff does necessary things that, although a volunteer might be inclined to do them initially, grow old over time."

NICs represent a general framework for performing knowledge-based activities. For example, the ideas in this book can also be translated into Engelbart's model. The first level represents the activities of champions and their teams, developing new innovations within a market segment. The second-level activity can be seen as the specific tools and processes we have outlined to speed up the innovation process, such as Watering Holes, NABCs, and Value Factor Analysis. The third level includes additional basic research into the Five Disciplines to develop even better innovation best practices. The facilitator at each Watering Hole meeting is the interpreter. And value propositions, innovation plans, and other documents used by the Watering Hole team represent content in the knowledge repository.

In Watering Holes, the goal is the rapid creation of high-value innovations. But a second goal—a meta-goal—is to develop methodologies that result in creating even higher-value innovations faster. As Engelbart has said, "The better we get, the better we get at getting better." If you want your organization to survive, that is superb advice.

We recommend that the interested reader learn about Engelbart and the Bootstrap Institute at www.bootstrap.org.

Team Dynamics

New ideas come from the genius of teams enabled by active communication. As Fusco pointed out, managing all the possible ways that team members can communicate with one another is challenging, because communication can take many forms: one on one, in subgroups, or among the entire group.

The law of exponential interconnections applies to teams, as well as to the Internet. For example, if each person in a team of five contributes and relates to the others as a collaborator in all possible subgroups, the result is seventy-five separate connections. You have seventy-five different ways for team members to share information, and each of these connections can have value. Twenty of the connections are to individuals—one-to-one. Fifty-five are with subteams and the full team. And each team member can have numerous connections outside the group. It is these relationships between team members that provide access to the collective intelligence of the team. This can result in teams with genius capabilities.

Each member of a community who shares ideas is supporting potential exponential improvement. But the example above, with only five members, illustrates a potential dilemma with teams: As the team size grows, the number of possible relationships also grows exponentially. When issues inevitably arise between team members, someone must have the skills and responsibility to resolve those issues. That is why most reporting relationships are limited to ten or fewer. Even in the Army, where the tasks are fairly circumscribed, the smallest unit, a squad, has only seven to twelve members.

When a team member begins a conflict, derails the group, or refuses to join in the collective vision, that person's power to drag down the community is magnified and can undo the potential value of many. A single person not only can prevent exponential compounding of ideas but can also create a negative exponential effect if others join in. You can spiral down just as fast as you can spiral up. One bad actor in a group can bring a team to a dead stop. In the exponential economy team members with both great skills *and* human values are required.

Human relationships are not free. They must be developed and nurtured to have value. All the connections between teammates also create the potential for confusion and misinformation. It takes continuous, respectful communication from the champion to keep the team together and focused on the task. A general rule for new champions, which is only a slight exaggeration, is that you should communicate with your team ten times more than you might have originally thought necessary.[7] This is because team members, like the proverbial blind men, are touching only a few parts of the elephant at any one time. You need to keep describing the entire elephant.

Your Innovation Teams

Obviously, just throwing people together on a team, whether face-to-face or virtual, will not automatically result in exponential improvement. Many organizations have team meetings that run all day long but that do not have a disciplined process to capture their genius. To leverage the genius of teams, you must have *disciplined processes* led by a champion. Many of us are often asked to attend meetings without any preparation or agreed-upon processes, or that fail to result in next steps. A colleague of ours often says, "A meeting without action items is a social event," which is not a productive use of people's time in an enterprise. As a result, many people just roll their eyes when they are asked to attend yet another meeting. If you cannot list the processes your teams use to leverage the unlocked genius of the team, it will stay locked. Do your teams have the requirements in place to produce exponential results? Do your teams:

- focus on an important customer and market need?
- have a multidisciplinary composition, where each person has a unique, complementary role?

CURTIS R. CARLSON AND WILLIAM W. WILMOT

- have passionate champions?
- iterate and compound ideas using shared language, tools, and processes?
- use continuous feedback to enhance contributions?
- share in the recognition and rewards?

Many people in organizations today find teams a drag rather than a value builder. When you sit in the corner and watch your teams operate, you can compare them with the dynamics described above and see where they can be improved. Do your teams activate $1 + 1 = 0$ rather than $1 + 1 = 3$ or more? If so, you need to invoke disciplined value-creation processes to let the genius of your team arise.

Often the biggest mistake a team makes is that the problem being addressed is interesting but not important. The cost of forming teams to solve small, interesting problems can be higher than the value that comes out of the solutions. Team formation takes time and brings costs and hassles. The rewards and benefits that flow from a project must be large enough to overcome these costs. People on teams can tell if you are wasting their time, and if you are they will stop attending, derail the team, or actively work to undo what others are trying to achieve.

Increasingly, businesses have "virtual teams" working around the world. If you are in New York at 10 P.M., for example, you may be on the Internet with others in Singapore, Great Britain, and India. Just being on the team—sitting through a conversation and doing e-mail while others are talking—is not a productive use of people's time. To harness the team's genius, each person must be engaged, contributing, and building on others' ideas in a disciplined manner.

You do not have to settle for suboptimal teams. Engelbart and his team are prototypes for many of the ingredients of an innovation team, but as Engelbart pointed out, it still took enormous effort. One of the hardest things to realize and teach is that success comes only when you deserve to win—and not one second earlier. That means that your team can't just have a meeting, brainstorm for two hours, and consider itself a high-performance innovation team. Team players will have to experience many failures and do dozens of additional iterations beyond what they expected when they started. Difficult projects are full of long hours and disappointments. They are mountains with a hundred false peaks. Having some buddies along on the climb makes the going a lot easier.

But when you are done, you have folks with whom to celebrate and share the rewards. When you and your team are done, as Engelbart's team illustrates, the results and rewards can be greater than you could ever have imagined. Teams can accomplish what seems impossible at first.

forming the INNOVATION team:

how we won an emmy for HDtv

FORM THE SMALLEST POSSIBLE TEAM,
BUT NO SMALLER. SEND ROSES.

Mission Impossible

Innovations go through many stages, from concept to prototype to product. As the program manager for developing the high-definition television (HDTV) prototype innovation at SRI's subsidiary, Sarnoff, Norm Goldsmith had his work cut out. We* needed to develop a full working model of our system in just over one year. It had to be delivered to the Federal Communications Commission (FCC) for rigorous testing in Washington, D.C., on a specific date. If we didn't show up on time, we would be disqualified. Tens of millions of dollars of research and development would be wasted. Billions of dollars of future business were at stake. Both corporate and personal pride ran deep in the competition to have the winning entry in the race to establish the U.S. standard for high-definition television.

The problem Norm faced was that our team *knew* it would take a year and a half to build a system of the magnitude that was required, not the thirteen months we had. As a result, we were facing a mutiny

---

* Coauthor Curtis R. Carlson was a member of the HDTV team.

by the team. Everyone was saying how completely crazy it was to even pretend we could finish the prototype. In fact, many thought it was foolish even to think that an all-digital approach to transmit HDTV would work in the first place, much less complete this prototype innovation in the time available. In addition, to succeed we would have to face the unknown difficulties of pulling together engineering teams from four other companies.[1] It was clear we should just face reality and give up.

In spite of the pessimism all around him, Norm went to work putting together a detailed innovation plan. With Glenn Reitmeier and Terry Smith, our terrific overall system architects, they interviewed everyone and helped each person think through their subtask to make sure the right people would work on them. When they were done they called the team of sixty engineers together in the auditorium at Sarnoff in Princeton, New Jersey. Covering the entire front of the room was a detailed computer printout of the innovation plan. It was a huge task. On the printout was every task that had to be completed, how each task interacted with the others, when each task had to be done, and the name of the person responsible for each task. A bright red line down the middle of the chart represented the "critical path." If you were on the critical path and you were late, the project was late too and you would get lots of attention. No one wanted to be on the critical path and be late!

As the team entered the room, you could feel everyone's pessimism, if not outright enmity. People had their arms folded in front of them and they had that "you must be kidding" look on their faces. Norm and Glenn stood up and began the presentation. Glenn started by giving an overview of our progress up to that point. It was a technical pep talk to help convince everyone that it was possible because our computer simulations showed it would work, we had built lower-performance parts of the system before, and at no point did it violate the laws of physics. Glenn made a compelling case. That was fine, but everyone still knew there wasn't enough time to build it.

Next Norm calmly described the printout and how it was constructed. He reviewed how he and Glenn had interviewed everyone. He said that he had taken everyone's plan as proposed but with a few exceptions. You could feel the team members say to themselves, "Exceptions?" Then he said, "Bruce, I think you were too aggressive in your plan, so I gave you another week. And Charlie, you need to integrate several other people's work. That is hard. I gave you two extra weeks." Norm, Glenn, and Terry went around the room going through the tasks, one at a time. Not once did they shorten anyone's plan.[2] They also explained how our terrific colleagues at Philips would build several subsystems. When he was done, he pointed out that the plan showed we would miss our deadline by just two weeks. You could feel the amazement in the room. We were close.

Then he said, "What do you think? If we work together can we find a way to make up those two weeks and deliver the system on time?" Much to their amazement, the team cautiously nodded their heads yes.

Norm and Glenn then explained how we would make up the time. A new, final plan would be created in which the most critical elements would be identified and then resources would be moved to those tasks as they showed up on the critical path. That is, it was the same plan, but Glenn, Terry, and Norm would strategically move a few staff as required and available to stay on plan. When someone needed to take over the system to test their part, they would be King for a Day, and they would wear a Burger King crown to show they were in charge. Glenn had the remarkable ability to make sure all the parts would mesh. As the system architect he understood the overall details of the plan.

That was the beginning of an incredible journey. Everyone had a critical role to play. We set up an isolated development lab in a building on the edge of our campus—a "skunk works." We brought in food three times a day and set up cots for sleeping. The team worked around the clock, eventually ramping up to three overlapping shifts a day, seven days a week. People were working 70- and 80-hour weeks, but they were also achieving the near impossible. It was a classic *The Soul of a New Machine*[3] setting.

Norm's project plan was on the wall so everyone could see how we were doing. The team performed amazingly, and we were managing to stay up-to-date on the critical path. But then, several weeks before the deadline and with victory in sight, the equipment failed to work. We had fallen off the critical path.

Our prototype system filled twelve refrigerator-sized racks of digital circuits and equipment. It was a monster composed of miles of wire and thousands of chips. Looking at it, you wouldn't believe it ever could work. It didn't.

Figure 12.1: Some of the members of the HDTV team at Sarnoff, SRI's subsidiary, in front of the equipment comprising the prototype HDTV innovation. Engineers from eventually nine different companies forged an effective team to help create the final HDTV system for the United States.[4]

The system had three major parts: an encoder, a transmitter, and a receiver. The encoder and transmitter would ultimately be at a television station. The function of the TV station encoder was to compress the high-definition video image by making billions of calculations each second to reduce the video's data rate by fifty times so it could be sent as a signal by the transmitter over the air to homes. At the home, an antenna would pick up the signal and then the HDTV receiver would decompress the signal and reconstitute it so it could be displayed on a high-definition TV screen.

A unique attribute of digital equipment is that if anything fails, the entire system may fail. That was our problem. We would put a signal into the transmitter, send it over the air, and then the receiver would display the signal on the screen. But the screen only showed scrambled lines and edges—no images. The team worked feverishly to find what in the system was broken. But it stayed broken.

Our schedule slipped, but fortunately so did our deadline—the competing teams and the testing labs were having their problems too. But even with the delayed deadline, the system stayed broken until the final few weeks. We were beside ourselves. Finally, someone checked a single connection and replaced it.[5] Miraculously, the system started to work, but not perfectly—it still had many periodic hiccups that had to be cured. The remaining few weeks had the entire team literally hunting for needles in a haystack—looking for a single digital bit that was wrong out of hundreds of millions that were fine. The debugging was so intense that planning was an hour-by-hour exercise. Glenn meticulously kept track of the debugging efforts on a huge whiteboard where the entire team could see it. No one ever needed to ask, "How are we doing?" It only took a glance at the board to know the answer.

The problems were fixed one by one, and eventually the system was "good enough" to ship to the testing lab in Washington, D.C. The prototype system continued to work over the next few months during the FCC tests, and it achieved the best picture quality among the competing systems. The team had achieved an amazing result.

Without Norm, Glenn, and Terry, this project would have been a failure. They were all champions for their parts and they made sure that everyone was in the proper job, that the right connections were made between the team members, and that there was a clear plan for everyone to see. At all times, they were respectful of their teammates. Norm could have come in with a plan that made the schedule. He did not. He depended on the genius of the team to help make the right decisions by having them be fully involved. Norm, Glenn, Terry, and many others[6] formed a high-performance innovation team and succeeded. Eventually, key parts of the system were included in the final HDTV system for the United States and the team won an Emmy Award, the highest award in the nation for broadcasting excellence.

In developing a successful innovation, there are several phases that teams must go through to achieve success. The first phase is to develop the value propositions for both the customer and enterprise and then the detailed innovation plan. The second phase is often to build a working prototype, as described above for HDTV. The third is to create the final product and deliver it into the marketplace. These different phases may involve different teams with different skills, but the fundamental human ingredients at each phase are always the same.

The Three-Legged Stool of Collaboration

Collaboration provides fuel for the innovation engine. When the members of a team really "click," they can win. The HDTV story exemplifies many of the attributes of innovation teams. The way the motivational issues were addressed was as important as the system's design. Failure to deal with them successfully would have resulted in project failure. This one vignette out of a ten-year project also illustrates that inventions are not innovations. The key inventions were not the issue at this point. The challenge was to deliver a working prototype to our first customer, the FCC.

Just about everyone has either watched or been part of a high-performance team. The New England Patriots, for example, pick players who fit together rather than looking for the star player. And they won the Super Bowl three out of four years, from 2001 to 2005. On the other hand, the Los Angeles Lakers under Phil Jackson in the early 2000s had some of the most individually talented players in the NBA, but could not master teamwork and didn't win.

One of our colleagues coaches a basketball team for eleven-to-thirteen-year-old girls. Every year, the girls play a boys' team for fun. The girls always win. Why? Yes, the boys are stronger and faster. But the girls pass the ball and play as a team. At the end of the game, the boys can't figure out what happened to them. Someday they will realize they learned an important lesson about collaboration from those girls.

Collaboration is powerful. Only through collaboration can one gather the skills and knowledge needed to solve most important problems and make an impact. Most of us enjoy working on collaborative projects with supportive colleagues. We want to make a difference and have our contributions valued. These are strong human needs.

Collaboration is possible when its essential elements are in place. Imagine a three-legged stool where written on the seat is "collaboration" and on each leg is one of these three elements:

- shared strategic vision
- unique, complementary skills
- shared rewards

Jerome Barnum, the inspirational founder of the Experience Compression Laboratory,[7] called this a three-legged stool because if one leg is missing the stool will fall over and collaboration will stop. The "stool" is held together by constant, respectful communication.

The Three-Legged Stool of Collaboration is fundamental to team formation. People form partnerships and collaborate *only* when all three elements are present. First, you must be able to understand and agree with the vision, goals, and objectives of the project. Second, you must be able to see clearly how your contribution is unique and essential to the success of the project. If your skills are redundant with those of others, then you are constantly worried about your role in the endeavor. And third, you must be able to articulate clearly how you will be rewarded fairly as a member of the team.

But these three ingredients are not satisfied automatically. Constant, respectful communication is needed to keep the Three-Legged Stool of Collaboration intact. Respectful communication is the glue that holds the three legs together. Without it, the stool falls over.

Norm, Glenn, and Terry and their HDTV innovation team were a perfect example of the Three-Legged Stool of Collaboration. If they had not shared the same strategic vision, and ensured that team members had unique, complementary skills and shared the rewards, they would have failed. The HDTV team remains the best example of managing an innovation at the prototype stage we have seen.

Shared Strategic Vision

A clear, compelling *vision* is a force for change—it pulls us forward. It is an image of the project or initiative when it has been successfully achieved. For a champion to gather a team and inspire everyone to achieve the objectives, there must be a clear, uniting vision—often with a higher purpose. As Proverbs 29:18 in the Bible says, "Where there is no vision, the people perish."

Doug Engelbart's vision was to find ways for people to work together to "make this world a better place using computer systems that significantly augment and extend human intelligence." For the HDTV team, it was to "reinvent television for the digital era by creating a profoundly improved television viewing experience for consumers, set the U.S. standard for digital HDTV, and form the core for video communication for the next century."

In addition to describing success for the initiative, the team's vision must meet four criteria to help propel the organization along a path of successful innovation. The vision must be:

- aligned with the organization's mission and goals
- clear, compelling, and forceful
- easily stated
- inclusive of all team members

You cannot have an innovative enterprise if the visions of some of your teams are not aligned with the organization's overall mission and vision. Each team's vision must also be clear, compelling, and forceful. A Dilbert-inspired example[8] doesn't quite meet this standard: "To be a leading worldwide innovator, producer, and distributor of state-of-the-art electromechanical subsystems with customer focus, moving value-based solutions to the market with maximum speed and high customer satisfaction, limited guaranteed performance, and online customer service for the good of the consumer, our employees, our stockholders, and humankind."

An example of a clear vision comes from DIVA Systems, which provided video-on-demand over cable TV networks. We asked Paul Cook, DIVA's chairman, to describe the vision for Diva. He said, "To provide immediate access in your home to all available movies at the same cost you rent them for today with no late fees and with all the functions of your VCR." Unlimited choice, lower cost, convenience, hassle-free. Sign us up!

A striking vision can stimulate interest. When a member of the Collaboration Institute, a consulting group, says, "Transform the workplace through collaboration to create innovation teams," it informs you of its core vision. You can easily formulate questions to ask, such as "How does collaboration work?" or "What is the nature of high-performance innovation teams?" Someone who asks an associate of the Collaboration

Institute, "How do you carry that out?" gets the institute's guiding mission:

- highest professional competence
- absolute truthfulness
- transformation focus
- custom services
- guaranteed quality[9]

These principles allow the associates to explain their vision in more detail. For example, customers are often intrigued by what "guaranteed quality" means. Guarantees are rarely given in consulting businesses, but the answer is very simple. "You are the customer, and you get to decide whether you are satisfied. If for any reason you are not satisfied, you get your money back." No client has ever asked for money back, in part because all the members are committed to the group's vision and the guarantee ensures that the members of the Institute will live up to the commitments made under their guiding principles.

Money is not a vision. When you talk to successful innovators, they rarely mention money; their sights are aimed at a higher purpose—a more important and transcendental level. Like the HDTV team, they want to make a difference. Other examples: "Reduce patient recovery by five times using minimally invasive surgery." "Construct a world-wide online education system for children with cancer."[10] "Develop the next-generation computer voice-recognition interface to the Web." When people focus on money rather than on the innovation for which there is a passion, they have usually lost sight of the original vision. Most innovators we have known who focused primarily on money have had limited success. Financial rewards are facilitated by a strong vision, not the other way around.

Unique, Complementary Skills

As Jim Collins noted, "Get the right people on the bus, the wrong people off the bus, and the right people in the right seats."[11] Only team members with unique, complementary skills and who can collaborate have a place on a high-performance innovation team. Every member must feel secure about having a significant role in the project. After all, you can't dance when someone is stepping on your toes. Ambiguity about the role of each team member prevents the commitment and collaboration needed for team success.

Innovation teams are powered by the collective intelligence of the members of the team. When team members are engaged in an iterative value-creation process, they can achieve customer value tens, hundreds, or thousands of times greater than each individual could achieve alone. But each team member must bring critical skills that are necessary for the success of the project. Each individual on the project must clearly understand his or her own role and its importance to the success of the project. The team members must also trust their teammates' skills, judgment, and determination to push through their parts of the innovation project.

As we noted in the last chapter, the powerful advantage of a team composed of people with complementary skills is that each member provides a different perspective.[12] That is why the collective IQ of a productive team can be at the genius level.

Although you should aggressively collect ideas from many, when developing the actual innovation, you want to form the smallest team possible. *The Mythical Man-Month* is a classic book on the difficulties of developing computer software in teams.[13] It explores why, paradoxically, adding more people to a software project can slow it down. You might think it would speed up productivity, since the lines of software code that can be written are proportional to the number of staff. But the coordination time between staff also grows as staff is added. At some point the advantage of more staff is overwhelmed by the rapidly increasing cost of team communication and coordination, which is an example of the law of exponential interconnections. The *smallest possible team* works best.

Once a group gets much beyond ten to twenty people, communicating and coordinating can become a champion's full-time job. If large numbers of employees are needed to accomplish a job, the only reasonable solution is to break the effort into logical subgroups to minimize the communication costs. The same principles we discuss here then apply to the larger group, as well as to each of the subgroups.

Forming teams takes time and effort—it is a project. It must be planned and managed. It is hard work. But a by-product of working with people with different skills and perspectives is the potential for an interesting and rewarding experience. Teams open up new intellectual worlds that can help you grow.

Shared Rewards

Rewards, like good meals, should be shared. In innovation teams, each team member expects to be rewarded for his or her contributions to the project.

Rewards come in many forms. A primary reward can be the opportunity to work on a terrific project with wonderful colleagues. In a company, rewards might also mean a plaque, a bonus, or a promotion. In a charitable organization, they might be recognition at a yearly banquet, highlighting the voluntary contributions of each team member. In a start-up company, they might be stock options. For the men and women of our armed services, the primary reward is the sense of achievement and camaraderie that results from performing an important job well. It is worth noting how proud military personnel are of a simple medal that symbolizes a moment of sacrifice or honor. Often a heartfelt "thank you" or "good job" is the best reward. All forms of recognition are important.

Rewards can also say a great deal about what the organization values. At SRI International there is an inspiring honor called the Mimi Award. It recognizes outstanding individuals who mentor and help further the professional development of others, which is a vital role in all enterprises. The Mimi Award is named in honor of the late Marian Stearns, who brought out the best in all the people who were fortunate enough to have worked with her. She would write notes of encouragement saying, "You are the best." She helped make sure it was true.

Rewards must be based on real achievement. A challenge for champions is to subordinate individual achievement and recognition to the overall success of the project. Often that means that if the project is not successful, there are no rewards for any member of the team—"all for one and one for all." This is often seen clearly on sports teams, where individual superstars give up individual achievement goals, develop and nurture a team identity, and win championships. At one point, the Boston Celtics, Green Bay Packers, Montreal Canadiens, and New England Patriots were all examples worth studying. For M. Edwards Deming, the prophet of the quality movement, teamwork and cooperation were effectively a religion. It was simple to Deming: *nothing* of significance can be done in a timely fashion by yourself.

Although it is common to celebrate success, there are times when it is even more important to acknowledge disappointments. Often bad things happen in business or life even though all the people on the team did their best. It is at these times, when everyone is most disappointed, that recognition of all the hard work is most appreciated. It also makes it easier to get up the next morning and start over. Moreover, in our experience, when we are working on an important problem, "the customer need does not go away." Most often these defeats are bumps on the road to eventual success and more work is required to find the solution. We need to keep our eyes on the eventual goal.

Managing people's expectations about their relative contributions can be a challenge for champions. It is human nature to overvalue your own contribution and undervalue the contributions of others. We tend to do this because we know firsthand what we did and how hard we worked, but we have only secondhand knowledge of others' contributions. Our colleague Carmen Catanese says that if you could give out 200 percent of the credit, everyone would be happy.

One way, as we have said earlier, is to focus on important problems, not just interesting ones, where the outcome is much greater than the sum of all the team members' contributions. And with great achievement comes the ability to give out more than 200 percent of the credit. If you have ever experienced what it feels like to be on a championship sports team of any type, you will know that this is true. Our success in helping create the U.S. standard for HDTV provided enormous rewards for all of us on the team.

There are other ways to make team members feel fully rewarded while you are working toward your goal. For example, we had the privilege of working with Sam Grant, a program manager for the U.S. government. Sam is one of those rare individuals who manage—through insight, determination, courage, and prodigious energy—to overcome the daunting difficulties of working within a bureaucratic government and to make major contributions to the United States. Sam is an energy emitter, not an energy absorber. He often said, "An attractive feature of credit is that there is an infinite amount that can be given away." He does that. Emulate Sam.

The DNA of Change

Change is hard. You have probably heard the quip "Change happens. Get over it." Well, yes, but only up to a point. The HDTV example shows how hard change is even for a superb, highly motivated team. All innovation work entails change, but real personal and team change happen only when three fundamental building blocks are in place. People change when they have:

- **Desire:** a need to change
- **New vision:** an acceptable place to go
- **Action plan:** a way to get to the new vision

Just as DNA is the fundamental building block of life, these elements specify the basic architecture for change.

Unless an individual or organization desires to move from the current situation, no matter how unattractive that situation might be, that individual or organization will not change. If people understand that there is a need to change but have no clear vision of where to go, then change is impossible. And assuming a convincing need to change and a compelling vision, change is possible only when a plan to go from the current circumstance to the new vision has been accepted.

If you have introduced an innovation, played a sport, or performed on a musical instrument, you have already seen the DNA of Change in action. Take the first-time skier, Stan. He falls down on every turn, spending most of the time sitting on the snow with a cold bottom and a *desire* (D) to do better. When he looks up the mountain and sees his friend having fun, whooping and hollering "Yahoo," he glimpses a *new vision* (N). He too would like to have the thrill of competent skiing. He hires a ski instructor who helps with the *action steps* (A), and the personal transformation begins. The coach says, "Stan, edge your skis, shift your weight, and face down the hill." As Stan brings the skis under control, he gains confidence, and his skill level increases. Within days he is skiing down the very slopes he had looked at longingly, and others are saying, "How does he do that? I can barely stand up on my skis." Stan has experienced the DNA of Change.

Consider as another example the short history of HDTV described earlier. The team understood the need for change, they had some appreciation for the new vision to get there, but there was no action plan. Consequently, the team came into the meeting with their arms folded in front of them: they weren't going anywhere. Only when the new vision and action plan were spelled out did the HDTV team get on board.

The DNA of Change applies to individuals, teams, and entire organizations working to develop new innovations. In each case, there has to be agreement on the *desire* for change, the *new vision*, and the *action steps* necessary to get there. If you don't have all three, the change will be unsuccessful. When any one piece of DNA is missing, the person, team, or entire organization will be paralyzed and incapable of change.

Have you ever gone through a major professional change, such as a merger, a job relocation, a promotion, or a change in direction for your team? If you have, you know that people actually use the words that describe the DNA of Change. In the beginning, they say things like "I don't see the need for this change." But once a desire for change has been established, they will then say things like "Yes, we need to change, but what is the new vision?" Finally, they will say, "The vision is good, but how do we get there? What is the plan?" If you listen closely you can tell where you are in the change process and when it is time to move to the next phase of enrolling the team.

Continuous, Respectful Communication

Collaboration is not free. It takes time and effort to create a shared vision, to form an effective team, to communicate over and over with the team, and to manage all the personal interactions between team members. Because these activities take time and effort, the project must be important enough to justify the investment.

Continuous, respectful communication is the glue that holds the Three-Legged Stool of Collaboration together, giving it strength and resilience. If there is no glue, the parts of the stool separate and it falls to pieces. If there is continuous, respectful communication, the stool can be made sturdy and withstand lots of weight.

Maximizing the potential collective intelligence of a team requires continuous, respectful communication between team members. Because of the different skills and viewpoints of the team, it must be expected that there will be misunderstandings about concepts, words, expectations, and rewards. This is especially true now that we have teams with different cultures working together all over the world. The champion must take the responsibility for identifying and solving these communication problems.

Our colleague Judy Fusco runs the exciting Internet-based collaborative education project Tapped In, which we described in Chapter 11. We asked her what the most important elements were in making the project successful with the thousands of researchers involved. She said, "Respect and trust."

We said, "Yes, that is true, but what else is important?"

She said a second time, with some emphasis, "Respect and trust are the most important elements in success."

We got her point. Fusco, as innovation champion propelling her team, spends a great deal of time helping mediate the relationships between team members and making sure that respect and trust are maintained.

When you value each member, work together to solve conflicts, and see them and yourself as tightly interwoven, you can engage in constructive criticism. You neither avoid difficulties that arise, nor do you ever threaten or coerce. You collaborate. You work out issues, move forward, meet team members' needs, and help them to be successful.[14]

The best results are achieved when difficulties are worked out *as soon as* you are aware of them.

Some of our colleagues who seem to manage the balance among individual support, high performance, and continuous team improvement are mothers with teenagers. There is a sense of "I care about you" affection with a practical, no-nonsense focus on outcome. The same approach works with teams.

Here is a quick checklist for you. It would be a good exercise to give copies of this short assessment to your team members, to see how each member rates the team on these crucial dimensions of team operation. On a scale of 1 to 10, with 1 being "not at all" and 10 being "totally," rate the following characteristics of your team. Write your numbers in the blanks. To what degree:

1. _____ Does each person share the vision?

2. _____ Does each person have a unique, complementary role?

3. _____ Are rewards shared?

4. _____ Are all members using respectful, collaborative communication?

Then ask your team members to individually fill out the scales. If individual scores vary widely or are not at least medium to high, you do not yet have a high-performance innovation team.

Support

No worthwhile project or undertaking can be characterized as "easy." People who are stepping up to a difficult challenge need a great deal of support from others. Our HDTV team moved into an isolated building to focus totally on the project. But the team members knew that they had the support of all of the companies involved. If someone from anywhere in any of the five partner companies could help the project, that person was made available at a moment's notice. Colleagues and supervisors understood that members of the HDTV team wouldn't be attending other meetings or working on anything else. The support of the entire organization was critical to success.

Perhaps most important, under the extreme conditions of the HDTV project we tried to be sure that all of the spouses and families understood our vision and the exciting mission for which their loved ones were sacrificing. Whenever a spouse or family member called an engineer on the HDTV project, we would ask to speak with them—to say thank you for their sacrifice, support, and understanding. The message sometimes wore a bit thin, but we never stopped saying thank you. On one occasion, we arranged to have a wife fly in to spend a weekend with her "estranged" husband. That expense wasn't in our budget, and it was against official company policy, but our terrific CEO at the time, Jim Carnes, agreed, and we did the right thing. The project fared far better than if our engineer had traveled home. The couple remained happily married, and when we finally shipped our HDTV system, we ordered sixty dozen red roses for all the spouses.

CHAPTER 13

OVERCOMING BLOCKAGES TO INNOVATION:

jim torpedoes a splendid idea

ADDRESS ALL TEAM CONCERNS IMMEDIATELY
AND DIRECTLY — WITH RESPECT.

Zero to a Thousand

Innovation is an intensely collaborative activity. While a lone inventor can wake up one morning with a brilliant idea, even incremental innovations take time to implement. The journey from initial idea to the marketplace includes many complementary activities performed by many skilled individuals over many months or years. In the case of major innovations, the journey can take over a decade, as illustrated by King Gillette and the disposable shaver, John Kelly's atmospheric radar, or HDTV.

Steve Jobs, during the middle of the Internet bubble, noted, "It's hard to tell with these Internet start-ups if they're really interested in building companies or if they're just interested in the money. I can tell you, though: If they don't really want to build a company, they won't luck into it. That's because it's so hard that if you don't have a passion, you'll give up. There were times in the first two years when we could have given up and sold Apple, and it probably would've died. . . . [T]here are many moments that are filled with despair and agony, when you have to fire people and cancel things and deal with very difficult situations. That's when you find out who you are and what your values are."[1]

The idea that a valuable innovation can be formed in a few months is an illusion. Kazuhiko Nishi, who is considered the Steve Jobs of Japan, clarifies the difference between invention and innovation: "There are two types of creativity: the creativity of making zero to one, and the creativity of making one to a thousand."[2] Innovation, moving from one to a thousand, requires a long journey and most of the hard work is invisible at the beginning. Consequently, some inventors overestimate their initial contributions and underestimate the contributions that must be made over time by the team.

Jim was, unfortunately, a classic example of going from zero to one. He was a brilliant scientist with a deserved reputation for coming up with potentially world-changing inventions. He had an idea for a new means of delivering insulin for diabetes patients. Today many diabetes patients must inject insulin several times a day to regulate their glucose levels, which is inconvenient and painful for the patient. It is also difficult to administer the proper amount of insulin. Jim had come up with an idea for a patch that would be applied to the patient's arm and worn throughout the day. It would collect fluid from just below the skin level and continuously measure how much insulin was needed. Simultaneously, it would dispense the needed insulin through a special membrane pressed against the skin. It was a potentially splendid innovation.

For the project to succeed, Jim would need to be the champion for developing the technology within his team. But Jim had a mixed reputation as a team player, so there was initial concern about starting the initiative. To address this concern, before the project was started his supervisors talked to him about what was required and what it meant to be a champion supporting his teammates. He assured everyone he was excited about the project and committed to the team. When he wanted to be, Jim could be engaging, charming, and persuasive. Because of Jim's promise and the opportunity to address an important unmet market need with a revolutionary approach, his management thought that Jim's involvement was worth the risk. Shortly after, a company was formed with a large capitalization of $6 million.

The trouble began almost immediately. Jim started complaining that his ownership of the company was too small, even though he had been given a disproportionate share at the start as a good-faith gesture to him. He hoarded information and refused to involve his colleagues. Not surprisingly, progress was slow and the technology continued to fail to work. Jim refused to be a champion and take responsibility, and in addition, he was derisive when talking to teammates about the CEO of the company. He made it abundantly clear that he did not respect his team members and, further, that he saw little value in the management team that had invested $6 million in his project. He constantly went around his management, and consequently hundreds of hours of management time were spent trying to get Jim and the team aligned. Each attempt resulted in more failure and greater animosity from the team toward Jim. As the saga unfolded, he blamed everyone else for the failure of his technology—his fellow team members and his management. Jim became, over time, more and more irascible. At the same time, he continued to demand greater monetary rewards and a bigger share of the company. Jim's lack of cooperation and greed were torpedoing the project. An impasse had been reached: the project was failing and Jim would not alter his behavior. The new company made the hard decision to remove Jim and start over by bringing in a new technical team, which took the project in a new direction.

Unfortunately, the chaos that followed the removal of Jim and the insertion of the new team meant that approximately $5 million and months of time had been wasted. In addition, because of the new direction for the company the CEO was replaced and others lost their jobs. Jim may have been a brilliant inventor, but he was not an innovator who respected the invaluable contributions of the team. His lack of trust for his teammates, which was expressed as lack of respect and generosity, poisoned the project. What saved the company was a superior board and investors who remained committed to the vision of the company.

Just as in Jim's case, you can only create a valuable innovation by working collaboratively. As a result, the behaviors of all team members determine success or failure. At any point, the team can fail because of all-too-familiar human issues. In the exponential economy, speed of innovation is a critical element of success. If the team lacks trust, it cannot perform fast enough to succeed. Similarly, if team members are gripped with fear, uncertainty, and doubt, or are corrosively cynical, productivity suffers. In extreme cases, when the team behaviors become too dysfunctional, teammates will give up and walk out the door.

If you are a potential champion for an innovation, like Jim, first you must get your own behavior in line with collaborative approaches. Second, you must address all nonproductive behaviors by your team members. If you do not, regardless of the importance of the project, its potential impact, or its financial rewards, you will fail.

The Trust Commandments

Although we have already mentioned trust in our discussion, it can't be emphasized enough. Trust is at the heart of teamwork. When it is there, it is a tremendous motivator. But without the bridge of trusting relationships, team members cannot rely on one another to build innovations with the highest customer value. In almost every broken team, when you ask, "What is missing?" someone will invariably say, "It's just that I can't trust him." Trust is easy to break and hard to build.

In our HDTV project, there were many times when trust issues arose; almost everyone got into trouble at some point. But in every case, the senior members of the team, who knew everyone, reinforced the message that all members of the team were valuable and reliable. That is, they could be trusted to deliver their part of the project. Just as important, the team members were trusted to know when to ask for help and trusted to accept help from other team members. When issues came up, they were addressed, and we succeeded.

There are some specific behaviors that build trust—an essential ingredient for innovation teams. The social bonds that hold these groups together are *respect for others, integrity,* and *generosity of spirit.* Without these values, trust is impossible. Trust allows team members to freely collaborate and to build on one another's ideas.

We call these human behaviors the "Trust Commandments," because they are so important. Commandments provide solid frameworks so that people, projects, organizations, and countries will not break down and people will not get hurt. Commandments are the few inviolable rules that must be followed; they are nonnegotiable. Violation leads, without exception, to severe consequences. In the case of Jim, he failed and was removed from the project primarily because he violated the Trust Commandments.

Respect for Others

As part of a working group of about a dozen people, our job was to make program recommendations to the U.S. government. Very well-known people were in the group, including a Nobel Prize winner, and all had strong viewpoints. Since we were going to be together day and night for two weeks, we were a little apprehensive about whether everyone would get along given all the overachievers in the room. But much to our delight, when the project was over it had been one of the most enjoyable and productive experiences of our careers.

Our panel chairman was Bob Galvin, former CEO of Motorola, who is considered to be one of the best U.S. managers of the past century. At the end of the two weeks we approached Bob to congratulate and thank him. And since these things don't happen by accident, we asked him what he was thinking about while running our group. After pausing for a moment, he said quietly and with feeling, "I believe respect for others is the most important value." Imagine how many problems in the world would be solved if people simply treated one another with respect.

Respect for others is the starting point for trust and productive collaboration. Respect is a demonstration that you value the contributions of others. Respect is not a social nicety—it is a necessity. If you don't respect others, they will not work with you and contribute to your success. It is also a state of mind: a belief that others have value and that you will make the investment to understand and appreciate their contributions.

Respect for others is a requirement for meaningful, long-lasting relationships with one's customers too. We have several colleagues from India who always give excellent, professional presentations. After one such presentation, we asked a presenter, Kumar, why this was so. He said, "In India, a customer is like a guest in our home."

The idea of customers being "guests in our homes" is a beautiful description of the kind of respectful relationship one should have with customers and colleagues. It doesn't mean that you give up your individuality and it doesn't mean that you have to agree with everyone else's ideas. It simply means that, above all, you will treat others with respect and consideration—as if they were guests in your own home.

Absolute Integrity

The news is filled with stories about unethical people. But with notable exceptions acknowledged, in our experience the business world is the opposite. Extremely successful people, like Bob Galvin, almost always exhibit superior levels of honesty and integrity. This is not by accident: the business world demands it. In the business world, relationships are voluntary, and if people are not respectful and trustworthy, then their negative reputation becomes known and people avoid working with them.

We were once asked, "How do you know whether someone will exhibit integrity?" In every situation, the answer quickly becomes obvious. For example, when you are negotiating with someone who, you discover, does not follow through on agreements, is abusive, tries to trick you, pretends not to understand your legitimate business needs, or takes credit for someone else's contributions, these are all clear indicators that this person is going to be a poor partner. And how will anyone who treats you in a punitive way treat customers and employees? It is hard for people to change their core values.

If your business partners, employees, and customers cannot trust someone in authority, you have a prescription for disaster. In our experience with hundreds of business deals, we learned that every time we had a doubt about someone's integrity and pretended that it didn't matter, it eventually led to a disastrous business arrangement. Integrity is not negotiable.

People of doubtful integrity also lead groups down a spiral of doubt and uncertainty where each person's motives are scrutinized at every point. In the exponential economy, where we must be able to rely on and leverage the unique capabilities of our partners and employees, even a modest amount of distrust will prevent the collaborative compounding of ideas that is necessary to succeed.

Conversely, integrity is contagious. When respect and integrity are encouraged, a robust environment can be created where these values become the team's norm. The CEO at Sarnoff when HDTV was being developed was Jim Carnes. Jim's integrity permeated the organization and was one of the keys to Sarnoff's success. The HDTV innovation would not have happened without the trust that Jim instilled throughout our team and in our corporate partners.

Every second we face a choice. Are we going to grow together or deteriorate together? The situation is never stable. Continual reinvestment is required to preserve the integrity and values of the group. Every business opportunity is filled with extraordinary highs and lows and numerous opportunities to question the motives of others. It is necessary to be constantly building your relationship "bank account" with others so that when one of these difficult periods comes along, there is "money" in the bank to draw on.

We were once involved in a potentially large project that included several different teams that, for historical reasons, doubted one another's integrity and value. Dan and Larry led these two teams. It was obvious to anyone who knew them individually that they both had extraordinary capabilities and were individuals of absolute integrity. Success on the project would result in a program ten times larger than either of their groups could achieve alone. Nevertheless, the personal baggage that Dan and Larry brought to the project continued to cause interference. It took many hours of working together for each of them to appreciate the value and integrity of the other.

In the beginning, simple rules were set up to help facilitate the improvement of their relationship. If one brought up a concern to us about the other, both were immediately called into the room to resolve the issue on the spot. Neither was allowed to talk negatively about the other. Finally, after Dan and Larry went through a difficult client negotiation together, where both displayed the integrity and unique skills that each possessed, their personal issues dissolved. Dan and Larry won their project and they are now great friends and productive colleagues. Most situations are like Dan and Larry's. People do have the integrity and skills needed. But sometimes a process must be set up and managed to remove doubts and misperceptions. It is one of the responsibilities of a champion to manage these relationships within the team.

There are only three choices. When working with someone else on a team who is difficult, you must do one of the following:

1.   Talk with the team member and work out the concern.

2.   Resolve it inside yourself, so that the other person no longer vexes you.

3.   Leave.

You must not form coalitions, criticize others in their absence, make end runs to the team member's manager, or use power plays. Work it out, let it go, or leave. If you can do it alone, do it. If you need help to make this happen, get help. The last option, leaving, is generally a poor one. All but a small percentage of cases can be successfully resolved if addressed directly and with respect for all the participants.

Generosity of Spirit

Greed kills. In the exponential economy, we are not individual operators. If you try to optimize your rewards at the expense of others, you will violate the Three-Legged Stool of Collaboration and create a disaster. We often ask people, "Would you rather have 100 percent of nothing or 1 percent of a million dollars?" Even though that sounds like a facetious choice, it is a fairly common example of what happens. We repeatedly encounter people who refuse to collaborate and share with others—and fail. For various reasons they choose a small, suboptimal return, rather than working with others to create a bigger opportunity. Innovation teams seek out and celebrate those larger achievements.

Do you operate from a scarcity model or an abundance model? If it is the former, you grab and hold all your goodies close to your chest, always guarding against others taking things from you. In the abundance model, you visualize yourself and others creating an endless expansion of rewards for all. And that is what you probably do.

Unfortunately, we have seen many individuals who either did not appreciate the value of others or were so greedy that there was no room for anyone else at the table. Greedy people almost always fail.

With your teams, build a "trust protocol." Ask your team members, "What behaviors would show that each team member operates with respect for others, absolute integrity, and generosity of spirit?" Put these on a flip chart and you will have a list for each person to follow—a behavioral prescription for building the trust that we all want. Then, after distilling the list and making sure everyone agrees to it, construct a final draft and have each person sign the protocol and agree to follow it. You have, then, your communication road map for how people will treat one another.

Other issues can interfere with a team's productivity. Next we discuss several forms of resistance and then unacceptable behaviors.

Resistance to Change

Change brings resistance. There are familiar behaviors that are played out as people move to a new vision. They include skepticism and FUD—fear, uncertainty, doubt, and misperceptions and red herrings.

Skepticism

It is a rare employee who hasn't been yanked from one project to another, or been forced to go through some management philosophy du jour. Champions must expect a certain amount of skepticism and push-back that says, "We tried this before and it didn't work," or "I don't get it—why does this apply to me or my customers?" Such skeptics are essential; they keep you honest and help you identify the issues that need to be addressed.

Each of us brings beliefs and expectations that determine our behavior. In part, our skepticism comes from lack of experience and success in the new endeavor and with the new team. Skepticism says, "Well, it might be a good thing to do that, but how do I know it will work?" Or, "Can you prove to me that this will apply?" Skeptics are from Missouri—the Show Me state. Comments from skeptics are important cues for champions, because they define what must be done to fix the concerns. Champions must actively listen to the team's worries, manage the relationships, identify each person's contribution, and point out the benefits that result. Such active listening is hard, but essential, work. In the next section on FUD, a conceptual tool will be described— reframing—that will help you to understand the skepticism you hear so that it can be addressed.

Once people in a group have had a positive experience in adapting to a new innovation challenge, they respond more productively when the next one comes along. For example, when a football team gets a new, unproven coach, both the football team and the community can be quite skeptical. But as the team members work together, develop common goals, sort out the unique and complementary roles that each will play, iterate the plays over and over to build confidence, and then share the rewards of victory, the skepticism declines. Proven coaches, like Joe Paterno at Penn State, build large reservoirs of trust that can transcend a number of poor seasons.

Organizations are in many ways like sports teams. When someone new comes in to lead a company, there will often be skepticism, especially if the previous management did not engage employees or was not successful. Skeptical people are often hard to convince, because they usually have had previous negative experiences. As a champion, you have to stay the course, never waver, and always be truthful and supportive.

People watch what you do, not just what you say. Any discrepancy between your actions and your words will be magnified enormously. Your staff can detect the slightest variance. Ironically, the people who speak up give you the opportunity to further dedicate yourself to the task of being a trustworthy person. Their constant demands for proof force you to understand your position deeply and why it will lead to success. They are like Cato in the *Pink Panther* movies. He was always attacking his boss by surprise to keep him sharp. The skeptical ones can do that too.

The FUD Factor[3]

When faced with significant change, many team members are gripped with FUD—*f*ear, *u*ncertainty, and *d*oubt. As the change comes closer, fear, uncertainty, and doubt emerge. When FUD overtakes us, we say things like:

- "I can't work with her."
- "Doesn't this violate who we are?"
- "This won't work."
- "This isn't fair."
- "Didn't we already try this?"
- "Change, change, change—that's all we hear."

When FUD is present, we can feel it, hear it, and sense it. It means that people are frightened. They do not yet see their places in the innovative activity. They can't visualize its success, and they are unable to see how their contributions will be valued. People feel disconnected from their strengths and the new vision. Once they see how they can connect to the vision and have a place in it, they will be fine. But in the beginning it looks impossible. The HDTV story, for example, in Chapter 12 started with the team confronting FUD.

It helps to view resistance, such as skepticism and FUD, as *gifts*. If reframed properly, people are telling you exactly what you need to know to move forward and be successful. Concerns usually have a kernel of truth that must be understood and addressed. Sometimes the kernel is a misunderstanding, sometimes it is valid, and sometimes it is wrong. Regardless, addressing these concerns will help you move forward. When we get an unexpected gift, we should unwrap it, look at it closely, and see how it can be used.

As we always say in the middle of a catastrophe, "Remember, a crisis is a terrible thing to waste." We always laugh, but this is right. It is an opportunity to address a problem that was sabotaging the project.

When an initiative isn't working for someone, FUD presents itself. Only when people feel that they have a place in the initiative will they be productive. As a champion, your job is to accept grumblings as a natural—and potentially helpful—part of the dynamic that leads to success. Just like the holiday present from your Aunt Betty, you may not at first see a use for it. But as a champion, you look behind the surface behaviors and transform them into a helpful part of the equation. First, however, you must engage the FUD and reframe it.

Champions must identify FUD and transform it for success. When people are acting angry and irrational, they are telling you what is scary to them—if you will just listen to them. When resistance kicks in, the person is not able to articulate what the positive message is underneath all the resistance. A competent champion, however, can listen to these complaints with a special ear—one that understands the message underneath the fear. The first key skill for the champion is to listen to the complaints and reframe them.

Examine the FUD complaints below and see how they are interpreted or "reframed" by a champion who has an eye for seeing them as clues for how to move toward a solution. The person utters the complaint—the champion hears it as the reframe. As you read the complaint, identify the FUD expressed and then restate it in a way that shows understanding and a sincere desire to address it.

COMPLAINT	REFRAME
"I'm too busy."	"I'm out of energy; give me some help."
"I can't work with her."	"She is different; how do I relate?"
"Doesn't this violate who we are?"	"How do I plug into this?"
"This won't work."	"I can't visualize it."
"I'll fight you to the death on this one."	"I feel excluded from the vision."
"Why do we have to work in teams?"	"Show me how to collaborate."
"No way will this be fairly done."	"Will I be rewarded?"
"Change, change, change—that's all we hear."	"I don't know how to adapt."

Embedded within each expression of FUD is often a kernel of hope that even the person speaking may not recognize. Circle back around, help them articulate their needs, positively highlight what's blocking them, and work through the issues one by one. You don't do it by e-mail, position statements, or color posters on the wall about cooperation. You do it in conversation with the people who need it. In general, these issues don't suddenly disappear. But if they are addressed directly and in a supportive way, most will attenuate and eventually go away.

Bill Mularie had a high-level position with the U.S. government. He has always pioneered exciting new innovations along the lines described in this book. As a consequence, he had, on occasion, met with people who perceived his new initiatives as threats to their organization and programs. Sometimes, in meetings with a dozen people, individuals would aggressively attack him. They were scared and expressing FUD. But he never responded with anger. Rather, he understood that they were afraid. He acknowledged their concern by saying something like, "Let me repeat what you have said to make sure I understand it," or, "That is an interesting point. I can understand why you might feel that way, but my experience suggests a different approach. Let me see if I can explain my view." He is a master at reframing.

At all times, Mularie was respectful, even when the attacks against him were personal. And because his responses were both respectful and reasonable, he always impressed the others in the room. He was seen as the kind of person with whom they wanted to be associated. As a result, he always carried the day.

Just as is done when an NABC value proposition is being developed, responses to FUD must be iterated many times. Each time you practice responding to a concern of others, you need to sharpen your eye, listen more intently, and respond with more support and understanding—all the while with the aim of helping the person or group move through the FUD. The champion is always alert to outbursts, challenges, and passive resistance, knowing that meeting people forthrightly and in a respectful way is the only path to success. Sometimes you may have a situation that requires outside intervention. If so, bring in someone to help those who are stuck.

Two program managers, Sally and Julio, were continually at odds with each other. During team meetings, they would roll their eyes and show disapproval of the other team. Their respective vice presidents tried without success to alter these behaviors. After realizing that FUD was overwhelming Sally's and Julio's teams, and spreading like a virus throughout the organization, the vice presidents called for a mediator to help. The mediator was able to help Sally and Julio resolve their issues and also coach the vice presidents on how to help Sally and Julio the next time something bubbled up.

The mediator interviewed each person separately to understand each individual's issues, reframing their fears as they spoke to help clarify their real concerns. He then had them meet to share those reframed concerns and find a solution. Finally, he had them write up a one-page agreement, which they both signed and which was monitored by the vice presidents.[4] When FUD springs to life, address it and look for ways to improve your response the next time.

As you work with your teams, see FUD as an important message. Eruptions are normal and to be expected—be ready to reframe these concerns and help the person address the FUD that is being felt. Someone gripped by FUD needs help. Your help can loosen the grip.

Misperceptions and Red Herrings

Misperceptions can create barriers to change. When first introduced, every new concept can have different meanings to different groups. For example, when one talks about "teamwork," it is very easy to convey the sense that teamwork in itself is a goal. But in companies working to satisfy real customer needs, teamwork is *not* a goal. Teamwork is the *means* to solve important problems and create value for customers— value that could not be created by individuals alone. To help make this important distinction, we refer to productive teams or "innovation teams," not just "teamwork."

People also have misperceptions based on erroneous "facts." We sometimes refer to these misperceptions as "red herrings." A red herring is something that is widely believed but is factually untrue and takes the project off track.

For example, it might be believed that the company's costs are too high, when a factual examination proves that not to be the case. When this is pointed out and proven, the reaction is usually "Oh, okay." Or think of Gilda Radner on *Saturday Night Live* playing hard-of-hearing media critic Emily Litella. After her outrageous pronouncements were proven wrong, she would say after a long pause, "Never mind."[5] Every company has its own red herrings: too expensive; we don't have enough resources; we are in the wrong location; we can't hire the right people; we have too many (too few) people. Red herrings must be identified and eliminated.

At the same time, every organization has barriers that legitimately need to be changed. It is equally important to identify these and separate them from the red herrings. In both cases, the key is to identify them, name them, and make them known, so that they can be either eliminated as objections or identified as key strategic problems that must be solved.

Unacceptable Behaviors

There is a family of unacceptable behaviors that will derail even the best-planned innovation. You may have experienced some of them. They include cynicism, passive-aggressive resistance, criticism of team members to outsiders, and end runs.

Cynicism

When change is being introduced into an organization, skepticism and FUD can become cynicism. And cynicism can grow to be a corrosive force, preventing the organization from moving ahead. Employees will say, "We have seen this all before from corporate. These silly ideas don't work." And "Yeah, the business development guys will get rich, but what do we get?" Or, "We're like mushrooms. They feed us garbage and keep us in the dark." Cynicism is corrosive. Cynics question everyone's motives and the motives implied are always bad.

The origins of these cynical views must be identified, brought to the surface, and met head on. If you ignore the cynicism, it takes on its own life, propagating throughout the organization. Cynicism keeps people from making a commitment to change.

Countering cynicism demands an enthusiastic, thoughtful, uncompromising champion. With a shared vision and visible signs of success, cynicism loses its force—and the team will begin to grasp the new positive reality. When team members shift from criticizing others to focusing on what they can contribute, cynicism has been defeated and the group will thrive.

Ultimately, cynicism cannot be abided: it will destroy any innovative idea. Or as a colleague says, "The problem with cynics is that they are always right—their cynicism, if not addressed, will destroy even the best ideas." Cynics must be identified and, if they do not want to join the team once their legitimate concerns have been addressed, asked to leave. Cynics are the mass murderers of productive collaboration and innovation.

Take the case of Mark, an ophthalmologist. He was in medical practice along with eight others, and they were trying to decide whether to bring the first surgery center to town, which could provide great new customer value. Unfortunately, he could be counted on to be cynical about every new idea. When it became clear that a surgical center could be built to provide lower-cost surgery for the group's clients, Mark, in his typical naysayer role, announced, "It won't work; it's just a lot of hot-air PR. And besides, patients can't tell the difference anyway." As a partner, he was on the verge of vetoing the new opportunity for the entire partnership. But the partners all took on champion roles, which included getting outside help. With the cynicism addressed up front and the issues worked through, Mark slowly began to get on board. As the partners directly addressed his concerns, he felt valued and listened to. During the ongoing discussions, he mentioned that he was interested in a new area of expertise: laser surgery for correcting nearsightedness modeled after the work Russians had done for years. The partners all gave him an enthusiastic response, noting how he could perform such surgery in the proposed center that would be cheaper for the customer and also give more revenue to the partnership than doing the surgery in the hospital. Four years later, the surgical center is a resounding success and Mark performs laser surgery, generating more revenue than any other partner.

Cynicism can be addressed and transformed. But it should never be ignored or tolerated.

Passive-Aggressive Resistance

Passive-aggressive resistance is obvious to all who see it. Some people believe that if they say nothing, they are free to walk out of the room and disagree with the conclusions of their group. A typical example of passive-aggressive resistance is someone with folded arms, leaning back in a chair, and vividly showing you, "I'm not engaged." Other examples include not returning phone calls or answering e-mails. Some people are more explicit and roll their eyes upward when others are speaking. This is neither fair to the team nor intellectually honest. It is also disrespectful and destroys trust. Each of us has an obligation to speak up and defend our ideas and then, when the team has made a decision, to support the decision. When you see body language that says, "I don't agree," it should be faced immediately. When in doubt, go around the room and ask for commitment to the decision.

Obviously, not all decisions made by groups are correct, but sabotage always leads to failure. We all have a responsibility to do our homework, be persuasive in the group, and help the group make the best decisions. If you sabotage your team, you kill your team. If you really feel you want to destroy your team and not become committed to transforming it into a innovation team, you should leave.

Criticism Behind Someone's Back

Criticism behind someone's back is toxic. It is especially poisonous because it destroys the cohesion of your innovation team and lets others know that you do not face and resolve issues. You spread innuendo, make false accusations, never take responsibility for your part of the difficulty, and keep the team from moving forward. Vicious criticism is like a cancer. It erodes the cellular foundation of teams and stops them from transforming into the exponential realm. At the same time, you reduce your effectiveness and diminish your own reputation.

Criticism to outsiders is illusory; it looks, on the surface, like you are engaging an issue. But if you really cared about the people on your team, you would raise your concerns *with them* and work out solutions. If you can't do this alone, you should get help from a facilitator. You work with others rather than against them. Criticism of others outside of team meetings is simply not acceptable.

Criticism of others saps the strength and purpose of the group. Sometimes simple, direct means of intervention will mitigate the problem. For example, in past experiences with criticism, we have resorted to telling team members, "If anyone comes to us alone and criticizes someone else, both parties will be immediately called into the room, where we will sit down and work out the issue." Typically, this needs to be done only once.

It is never helpful to let "behind-the-back" criticisms go unchecked. You render yourself impotent, set distrust and disrespect in motion, and destroy productive collaboration.

End Runs

We have all experienced end runs. In grade school, we learned that tattletales were the worst kind of friends. You have a disagreement with Billy and call him a bad name, and he runs to the teacher saying, "Teacher, Jimmy called me a bad name." The "end run" is when you go around a person and take your criticism to their supervisor.

Not only children do end runs; some adults specialize in it. Have you ever been on a team with someone who was bothered by you and, instead of working out the issue with you, ran around you and went to your manager? Did that help build trust?

Recently, a new dean was hired at a large public university with a mandate to bring innovative approaches in teaching and research to his fine-arts school. Within a few months, it became clear that many of the fifty school faculty members didn't like their dean. Instead of meeting with him and working though their concerns, 85 percent of the faculty signed a petition asking the provost, the dean's superior, to fire him— a classic end run. But the provost flabbergasted the faculty by saying, "I am sorry you did this: the dean will stay. But we need to work through these concerns." Because of all the pain and disruption, the provost called in an outside mediator. After interviewing the faculty and the dean, and spending another half day with the entire faculty in a workshop, there was a four-hour dialogue with the dean, where the entire faculty and the dean met in an open forum. At this meeting, led by the mediator who set clear boundaries on how to engage issues in a productive way, the hottest issues were put on the table—and to rest. In fact, the dean and faculty set out a process to work together to address the remaining few concerns. This end run did get people's attention. But initially the faculty came off as whining, uncooperative, and unwilling to give the dean direct feedback about his behaviors that caused concern—a no-win situation. It was a huge waste of everyone's time.

In any serious team dispute, someone must be the mediator or facilitator to identify issues and put together a plan for resolution. The mediator or facilitator can be a trusted colleague, a supervisor from another unit, or an outsider. Getting the parties together and having them generate a communication protocol, a short contract outlining how they will behave, is particularly effective. Each party must sign the protocol or contract and agree to a follow-up plan to check on compliance. In serious cases, we recommend getting outside professional help. If your project or team members are at risk, it is a worthwhile investment.[6]

Those who do end runs often justify them by saying that it was the only way to secure change. This is usually self-justification for poor behavior, fear of conflict, and inadequate interpersonal skills. End runs inevitably result in loss of trust, which then takes prodigious effort to repair.

Criticism and end runs undermine the trust that must be established in all innovation teams. If you break confidences and say terrible things about team members, because you incorrectly believe you will look good to those who evaluate your performance, you are doing direct and collateral damage.

Do we need to write more about end runs and breaking confidences? Don't do them. Don't let others do them.

CHAPTER 14

INNOVATION motivators:
saving Larry's Life

"THE FUTURE BELONGS TO THOSE
WHO BELIEVE IN THE BEAUTY
OF THEIR DREAMS."[1]

Eleanor Roosevelt

Vision

Human tragedy can be a strong motivator for innovation. Our colleague Peter Burt and his team were inspired by the illness of a friend to take their work in computerized human vision to a higher level. Human vision is one of the most complex and impressive sensory systems in nature. The amount of computer processing needed to perform even the simplest vision tasks, which humans do effortlessly, exceeds the power of the world's biggest computers by factors of millions. For twenty years, Burt and his remarkable team have led advances in the development of real-time computer vision systems that rival the capabilities of human vision. They have created a wide array of significant innovations, from automobile traffic monitors to biometric identification systems, to airport security systems, to broadcasting equipment, to medical diagnostic systems. Each innovation was motivated by an important need.

One medical innovation was inspired by the team's desire to help save the life of Larry, who had an inoperable tumor next to the nerve in his spinal column. Conventional radiation therapy was not an option, because even small changes in Larry's position on the X-ray table would cause spinal nerve damage that might render him paraplegic. The only viable approach was to use Stanford University's newest development, the CyberKnife, which had a robotic arm that adjusted the aim of an X-ray source to compensate for patient movements. This system tracks the patient's movement using a vision system based on CT images and adjusts the position of the X-ray accordingly. It was determined, however, that Stanford's vision system needed to include both CT and MRI imaging to have the precision to treat Larry's tumor. This new precision vision system even had to take into account the articulation of the spine.

Larry had heard a talk, "The Age of Innovation," by one of our team, and learned of the highly precise targeting technology developed at Sarnoff. He contacted Burt's team, and the team decided that this was an important need. It swung into action.

As in every project, there were endless issues to be addressed, from the practical to the political. The team worked long hours for weeks to implement the solution. Championed by Teddy Kumar, Jim Bergen, and Norman Winarsky, the team was able to precisely locate the tumor for the Stanford CyberKnife team, which then successfully extended Larry's life without leaving him paralyzed.

Why do people work this hard? It's not for the money. We want to address higher needs in our professional lives, and Larry's situation was a compelling example. The opportunity to help extend someone's life presents a rare and special opportunity. But all good employees share the need to make an impact through their work.

The Motivation Mantra

Innovation is inspired by fundamental needs that motivate. Without the motivation in place to "stay the course," disruptions will occur that can prevent your innovation from reaching fruition. We call these three basic human needs the Motivation Mantra: *achievement, empowerment,* and *involvement.* Burt's team has always satisfied these motivators, which is one reason their impressive list of innovations continues to grow.

Achievement

People want to make a positive contribution in their careers. They want to make an impact and achieve meaningful goals. By focusing on important customer and market needs, your team can contribute and make a difference.

Paul Cook is an unrelenting serial entrepreneur. His first major company was Raychem, where he developed a family of novel plastic materials. He has won many honors, including the National Medal of Technology. When he retired from Raychem, he formed CellNet, and after CellNet he formed Diva Systems. And at seventy-five he formed another new company, AgileTV. Why do people who no longer need to keep working so hard? Clearly, they want to continue to use their skills to achieve goals that are important to them. This is an enormously powerful human motivator.

Consider our friend Carla, who works as a volunteer in a hospital for severely handicapped children. The children she works with are all born deaf and blind. To communicate, they tap out a form of Morse code on one another's hands. Progress is often painfully slow. In some cases, it takes years of training for a child to learn even the simplest messages. We asked Carla why she did this work and she said, "Oh, I love it. You can't imagine the satisfaction when a child, after years of work, first taps out a simple message on your hand." This is certainly a major achievement.

Similarly, we have a family member, Donna, who is a nurse. She works in a state psychiatric ward, where patients with the most extreme emotional and mental health issues are confined. The patients include schizophrenics and sociopaths. To an outsider, it seems like the most dangerous and demanding work. When asked why she does this when she could work in any hospital, she said, "They truly appreciate what I do. It's important."

Most people do not want to work just for money, stability, or enjoyable friends. We want to learn new skills, become more valuable, and be recognized and appreciated. We want to contribute by doing meaningful work and by making an impact. These objectives are, of course, especially possible when you are working on an important innovation.

Empowerment

We all want freedom to do our jobs. If your manager gets in your way, throws up roadblocks, and doesn't give you latitude to perform, you won't. One of the most common complaints of employees is that they are "micromanaged" and not empowered to perform.

When the Three-Legged Stool of Collaboration (shared vision, unique complementary skills, and shared rewards) is satisfied, the individuals involved need to be empowered to do their jobs. They want to be respected and make a contribution without being told *how* to perform each step. They want to be turned loose to experience the surge of creativity that leads to success. Talk to folks who do not feel empowered, and you hear the cry for help: "All I do is go to meetings," or "It is just impossible to get work done around here with my boss looking over my shoulder." Just as the economy works better when markets are free, people work more creatively when they are set free and empowered to do their jobs.

Champions work with innovation team members to define the "what" that must be done, and then turn them loose to figure out the "how." Support them and hold them accountable, but do not tell them how to do their jobs. Do more of the "what" and less of the "how." Empowerment means assuring the independence and respect for others that allow them to do their jobs as professionals.

Giving people the respect and freedom to do their jobs does not mean everyone is a free agent. Each person needs to be reviewed on a regular basis to make sure that that individual's project is on schedule and that all the elements needed for success are in place. Too often we see employees in companies with the responsibility for an important project but without the authority. Champions must, at a minimum, (1) pick the team, (2) review the team, and (3) reward the team. As management guru Peter Drucker points out, these are the minimum responsibilities when running a project.[2]

The review function is not only an opportunity to check on progress, which is essential, it is also a major opportunity for improvement and additional value creation. Regular constructive reviews are necessary to realign the team's vision, collect new ideas, and provide support and acknowledgment for progress made. When we are stuck and cannot figure out how to progress, these regular reviews provide a mechanism to support empowerment and move us forward.

Involvement

If you are leading any innovation, you are a change agent responsible for many decisions. We have a strong need to be involved in decisions that affect us. Not including members of your innovation team in decisions that affect them is extremely dangerous. Resistance to new initiatives may go underground for months or years, but eventually it will rear its head and impact the team's productivity. When people are not involved, they become angry and irrational, because exclusion directly threatens their ability to contribute—one of their most important needs.

In our intervention work in organizations experiencing severe disputes,[3] it is almost always the case that the person causing the ruckus was left out of an important decision: what work to perform, that person's role in it, or the new direction for the organization. Because achievement and empowerment are so important to all of us, not being involved threatens our very identity. Professionals especially, if they are not involved, will stop being solid contributors or even leave the company. Involve them or suffer.

In the early years of safety practices, it was thought that you could decree perfect safety practices and employees would follow them. Then it was found that, if asked, employees would develop the same safety practices as the experts.[4] The difference is that if you take the time to let employees develop the standards, they follow them rather than sabotage them. To get results for your innovation, trust your people and involve them.

Involving others in important decisions is also an indication of your respect for them and your belief that they have important insights that will contribute to better solutions. In our experience, that is almost always the result—a better solution.

Clearly, there are situations where decisions must be made that individuals will not like. Nevertheless, when people are properly involved, they will generally accept decisions that are in the best interest of their enterprise. We can give you a guarantee—if people are not involved, even if the decision is correct, you will end up doing remedial work to get the organization back on track. If you do involve them, even if the decision is different from the one they would like, you have a good chance to gain their support.

There are exceptions. Some organizations are hopelessly dysfunctional and uncompetitive. Occasionally, people will refuse to change. Others lack core skills and will, for a variety of reasons, consistently make terrible decisions. In these cases, it may be impossible to implement the needed actions. Fortunately, we have found these hopeless situations to be fairly rare. The major problem we see is that no one is continuously working to address the Motivation Mantra. Overwhelmingly, in situations where we have been called in to mediate an organizational conflict, from people screaming at one another to entire departments signing petitions to have the manager fired, someone has been ignored or not included in major decisions.

A situation that is particularly demanding is mergers and acquisitions. In these situations one company is combined with another while parts are closed down or sold off. Even here, involving all the staff fairly and honestly makes a significant difference. When RCA was bought by GE, we went through this process. Fortunately, GE was expert at these kinds of activities and did a good job.[5]

As a champion leading an innovation team, your job is to align all the elements of your team, keep them focused, and constantly work in the direction of change while valuing what you already have. If you rush in with change, pretending that employees are not the most crucial part of the organization, you will fail. Instead of coercing others, let them join with you to create additional value.

The involvement of others is so fundamental that if you violate it, you will almost guarantee failure. Time after time, change agents' failure can be predicted by seeing how they involve others. This is highlighted by the Failure Formula—a checklist of what change agents often do that brings disaster.

─────────────────{ THE FAILURE FORMULA }─────────────────

- Bring in a grand vision that others didn't help create.
- Make pronouncements about what needs to change without first talking to those who will be affected.
- Spend all your time with those above you, not with the rank-and-file employees.
- See and talk about only the positive aspects of "the change," never acknowledging possible downsides.
- Make big symbolic changes only—fire people, restructure units, change the company logo.
- See those resisting you as aberrant personality types, questioning their motives.
- Never give individuals a chance to discuss, modify, and work through elements of the plan.
- Talk only to your "supporters," avoiding those in the "other camp."
- When resistance arises, stop having meetings.

A very large international organization with which we worked was going through a major reorganization and layoff. The new president put together his plan and was about to announce his decisions. It was suggested that he might involve the members of his extended management team, because they would have to implement the plan. At first he demurred. But fortunately he agreed, and a session was held where he first described the need for change, his vision, and his proposed plan. He then *sincerely* asked for suggestions. He broke his management team into working groups around specific topics. The groups then reported out their thoughts. They too came to most of his conclusions, but they also had ideas that he agreed were better. In some cases, he still needed to make hard decisions in opposition to his management team. Nevertheless, the reaction by the management team to his involvement was overwhelmingly positive. As a result of the meeting, he had gone a long way toward getting his team aligned with the new needs of the enterprise.

Successful champions simultaneously listen to and lead their innovation teams. Since change is constant, champions must be alert to the sounds of discontent, air issues, and work them through.

Few innovation guarantees are as ironclad as this one. As a champion, to be successful, involve all the members of your team, collect their ideas, and gain their support.

Involvement also includes having a supportive community. Work in the exponential economy takes great effort. We all need others to hold out their hands, giving encouragement and helping a project move forward.

Regardless of our profession, we each work with colleagues and other professional groups. These groups are sources of inspiration and friendly competition that help us strive for greater achievement. Community is the gathering of others around us, giving us a sense of place and importance. If you are working on an innovation, convene a community of fellow innovators. It is a great feeling when you have a difficult problem and, after a colleague engages you in conversation, a solution arises.

Involvement in communities at work creates value. Regular group meetings, parties, and award ceremonies are all essential for creating a sense of community, recognition, and personal worth. In a healthy community, team members take care of one another. We have social scientists at SRI and they are particularly good at this. They bring food to share for breakfast, give out funny awards on special occasions, and have events like an annual chili competition, where everyone comes and celebrates. Softball teams, bike clubs, and other after-hours activities all help too.

Money is not on our list of key motivators. As long as compensation is fair and reasonable, we find that failure to satisfy the Motivation

Mantra is what most often gets employers in trouble. Interestingly, when surveys are done, 80 percent of managers believe their employees leave because of pay. When asked, 80 percent of employees say they leave because some element of the Motivation Mantra was missing.[6] That is our experience too.

We have described three basic human motivators: achievement, empowerment, and involvement. Check with your team and see how you are doing and perhaps, like Burt's team, they are on the path to becoming top performers too.

Steps of Personal and Team Transformation

Most of us get paralyzed in the face of change. The movement to something new requires crossing a deep void—a subconscious barrier. We perceive a gap between the old vision and the new vision, a bottomless pit into which we might fall. It appears because we are not sure that we will be successful and valued in the new vision or assignment.

At different times we all need someone to help motivate us to make a change. Even in the case of Burt's group tackling Larry's need, the team needed to cross the void. It was not obvious at the start that they would be successful, but it was obvious that it would take a prodigious effort.

The secret of crossing the void is deceptively simple. The secret: We must see a way to *leverage our current strengths in the new innovation's vision.* Some of us will be able to make this transformation on our own, but most of us need help from our significant others, colleagues, and other champions. As champion of an innovation project, helping all the members of your team to cross the void is your responsibility and one of your first priorities. You can't get up to full speed until everyone is across the void and fully on the team. Helping others means addressing the DNA of Change, satisfying the Three-Legged Stool of Collaboration, fulfilling the Motivation Mantra, and employing Personal Transformation to help them become fully engaged in the new vision.

Helping each team member move to the new vision follows a predictable sequence:

- Identify the strength to be leveraged.
- Expand and incorporate the strength into the new vision.
- Take small steps to try it out.
- Engage other team members to help.
- Give up elements of the old vision that no longer fit.
- Celebrate the results.

At some point, elements of the old vision must be left behind if the new vision is to be successful.

We see examples of the need to help others vault the void every day. Take, for example, the case of Kevin. Kevin comes from Ireland and is a lively, some would say exuberant, talker who manages a major new contract with a Japanese firm. He is exactly the kind of fellow one would enjoy talking to in an Irish pub. But he is just the opposite of his Japanese counterpart, Takeshi, who is rather quiet. Kevin and Takeshi were about one year into a complex business relationship when Takeshi sent an e-mail to Kevin saying, "The contract is canceled." Kevin was caught by surprise, so he constructed a long, negative e-mail in reply. But before clicking the send button, he decided to call a colleague and tell her that the contract was in jeopardy. The colleague recommended that Kevin: (1) not send the e-mail, (2) immediately schedule a trip to Japan to see Takeshi, (3) sit down with Takeshi to ask him privately about his needs, and (4) practice using listening skills.

Kevin said, "But I don't know how to get him to tell me what his real issues are. He doesn't talk much." "Kevin," his colleague said, "imagine that you are in an Irish pub, drinking with Takeshi. Pretend that you are talking to another Irishman, just like you did for years. The key is to listen to Takeshi and get him to talk to you. Make sure that he talks at least ten times as much you do. You have all the skills; you just need to move them into the new environment."

Kevin practiced with another colleague to prepare appropriate questions for Takeshi. During their meeting and after a few bottles of sake,

Takeshi finally got around to describing his real problems, which were mostly about politics in his own company rather than about the substance of the contract. Once Kevin understood Takeshi's needs, they were able to solve them together. The relationship was saved and the contract was preserved.

Kevin leveraged his strengths. He had the necessary conversational skills, which he used to great effect in a pub. They just needed to be identified and applied to his new situation with Takeshi. Each of us has strengths that can be leveraged to vault the void that separates us from the new vision.

Crossing the void is also hard because it challenges our *identity*. Innovations often require that we perform different jobs as the innovation progresses, which requires changes in our professional identity. One day we are a basic researcher, then we are a champion leading an innovation team, and then we are part of a business-development activity. Changing our identity is one of the most difficult things we can do. If a situation involves a significant change in the identity of an individual, team, or organization, you are embarked on a multiyear endeavor. It is a major project. You should treat it as such.

Here is a common example that directly affects many professionals: If you get an advanced degree at a university, you are expected to do original research—by yourself. Collaborating with others is discouraged and treated like cheating on an exam. When you leave the university, however, the most powerful way to work is with others in innovation teams. This team identity is different from the university individual-contributor identity. Working on an innovation that necessitates collaboration will challenge this individual-contributor identity you bring with you. Personal Transformation, when applied step-by-step, is the key to addressing this identity change.

Another common example is someone who is asked to champion a project for the first time. This new champion identity is different from the previous team member identity. It is also scary because the necessary skills are different. And the rewards that come from helping others succeed are uncertain at first.

One of us can remember the first time we were asked to lead a new program and we walked past the vice president's office two dozen times—trying to decide—before we accepted the assignment. These transitions can seem incredibly hard while they are going on. But once the change is accomplished, it is difficult, in retrospect, to remember why they were such a big deal.

Once you cross the void to the new vision, you can serve as coach and assistant to other team members who are just joining the initiative. Pretty soon, the entire team is centered in the new vision, talking about being part of "a world-class team," "one of the best departments of communication in the country," or "the finest custom boat builder in the industry."

Great teachers intuitively understand Personal Transformation. Throughout grade school, we all experienced such transformations, but we probably didn't pay attention to the process. In elementary school, if we were good at spelling, the teacher leveraged that to help us understand sentence structure. If we could add numbers, that was used as the bridge to get us excited about multiplication. If we enjoyed a good story, that became the springboard for writing literature. If we were quick-witted, we became debaters and public speakers. If we enjoyed drawing, that was transformed into a career as a commercial artist.

Examine your own situation. First, see if your team satisfies the Motivational Mantra: achievement, empowerment, and involvement.[7] Then, if your team is not fully aligned with the new vision for your innovation, employ the steps of Personal Transformation to help them with the journey by leveraging their strengths.

DISCIPLINE

5

orɡanizational

aliɡnment

YOUR
INNOVatION team:
you can start now

Crisis

Back in 1986, my colleague Norman Winarsky and I were mid-level managers at RCA Laboratories in Princeton, New Jersey, responsible for a team of about seventy researchers.* That year GE bought RCA and shortly thereafter RCA Laboratories became part of SRI International, which created a separate enterprise called the Sarnoff Corporation. At RCA, each year we were given a budget for doing our research. Once we became a subsidiary of SRI, however, budget sources were neither automatic nor a given. We went from working for a corporate research laboratory to working for a research institute funded primarily by clients. We had to develop the business needed to support our people. This was a 180-degree turnaround. Unfortunately, like many managers of large companies, we knew little about customer needs, business develop-

---

*This section is in the voice of coauthor Curtis R. Carlson.

ment, and business models. We knew even less about how to create compelling customer value. We had a crisis on our hands. We did not know how to succeed. Only later did we discover that this would turn into the best opportunity of our professional lives.

Out of necessity, our team met every two weeks on Monday from 5 to 9 P.M. over pizza to share what we had learned. Everyone, including Norman and me, stood up and gave short presentations to the group about our business ideas. Then the team would critique them. For the first few years, our progress was painfully slow. But we all improved and learned from one another. We began to succeed.

We recognized that we had to understand and develop innovation best practices that we could apply to our new jobs. We collected, wrote down, and shared our best practices. Ad hoc presentations became NABC value propositions. Words like *iteration* and *compelling customer value* became part of our shared language. After a few years went by, significant results began to show. We were winning more and more important projects and growing rapidly. Others began to take notice. The successful innovation practices developed by our team began to spread to other parts of Sarnoff. Norman and I continued to explore new ways of creating value, and eventually spun out the first company from Sarnoff. Finally, at our team's insistence, we ended our Monday meetings earlier, at 8 P.M., so everyone could go home and watch *Monday Night Football.* Ending earlier was an important symbolic victory on our path to success.

Your Innovation Team

Our experience proves that innovation best practices can be applied locally within your team. They do not require upper management's permission. As you can see from our pizza-meeting story, the origins of these ideas came from a group that was midway down in a large company. There was no executive support or acknowledgment for these activities at the time. It was neither asked for nor needed.

There are only two concepts that must be embraced by your team to get started. The first is the goal of creating and delivering the highest customer value in the shortest time. The second is to use innovation best practices as the means to make that goal happen—the Five Disciplines of Innovation. Involve your team in this discussion and obtain their commitment.

In our search for innovation best practices, we have worked with many other colleagues who share our passion for creating the highest customer value, including Herman Gyr, a world expert at creating organization structures that allow enterprises to thrive. At SRI we now use these innovation best practices across all parts of the enterprise. Everyone—and we mean everyone—gets exposed to these ideas and uses elements of the Five Disciplines in their work.

Your team may reside anywhere in the enterprise. You could be in public relations, marketing, finance, engineering, or the CEO of the company. You can start with your team now. In the next chapter we will explain how to launch the Five Disciplines of Innovation throughout the entire enterprise.

We have outlined the Five Disciplines that, over time, need to be embraced and deployed throughout your innovation team. First, focus on important customer and market needs. If you have external customers, you must have a deep understanding of your customers' and market's ecosystem, as described in Chapter 9. If you have internal customers, you start by developing a deep understanding of their needs and how they aid in the creation of new customer value. Try a little test to see where you are. Within your team, have your team members write down who their customers are and what they think their customers' needs are, and then ask your customers to do the same. We predict that you will be amazed at how little overlap there is. We also guarantee that comparing the lists will trigger a valuable discussion.

Second, use the family of value-creation tools to create new customer value. The tools include NABC value propositions, Elevator Pitches, regularly convened Watering Holes, and full innovation plans. Along the way, determine metrics for important customer and market needs. Quantify customer value using Value Factor Analysis and other evaluation tools. And, over time, the Watering Holes will help your team members constantly improve their value propositions, share learning about business concepts, and gain greater skills.

Third, make sure that each innovation initiative has a champion, who steps up and keeps the team on track. You must make this person responsible for the overall initiative, so that when difficulties arise, the champion can take corrective action—whether that difficulty is motivating the team, clarifying the vision, synthesizing the solution, garnering financial support, or finding other champions for parts of the project.

Fourth, as the project comes together all the innovation team dynamics we have discussed will emerge. You must select the right team members, fully involve them, solve the inevitable concerns that arise, and move forward. The words they use will tell you where you are in the process and will help you identify issues to be addressed. You will literally hear comments like "We don't understand the need" and "This is all well and good, but it doesn't apply to us." As we have discussed, these are gifts that you should embrace to help you understand the issues that must be addressed, to make progress rapidly. Only if you have a motivated and dedicated innovation team can you have success.

Finally, fifth, you must align your team for success. Then, as you begin to have success, slowly educate other players in the broader enterprise so that customer value becomes the central hub around which all activities turn.

Team Alignment

A team cannot perform to its fullest potential unless it is aligned with the goals of your innovation and the innovation is aligned with the goals of the enterprise. Alignment means that barriers to success have been eliminated and the organizational support needed for success can be put in place. Team and innovation alignment starts by achieving the other four disciplines, as listed above. Failure to achieve alignment within the project, the team, and the enterprise can lead to either costly delays or outright failure. As barriers are identified in the project, team, or enterprise they should become part of your innovation plan with actions to address them.

Removing Barriers

Barriers to success come in many forms. For example, putting a new initiative that conflicts with an existing product in the same department can result in failure. The inevitable organizational conflicts slow down the project and frequently kill it. Often the best success scenario is to create a separate organization for the new initiative. For example, GM is developing fuel-cell power systems for vehicles that, if successful, will replace today's internal-combustion engines. Because of the potential business conflicts they, wisely, created a new organization that does not report to the existing engine organization.

Sometimes it is best to create an entirely new enterprise. At SRI International we incubate major initiatives within the organization and then, when we have developed a complete innovation plan and raised the resources needed, we spin them out as new companies. This is required because SRI is aligned around the initial stages of innovation, not mass product production and sales. Major innovations need their own dedicated team and infrastructure to create their specific products and services. In a similar way, universities establish centers and institutes to ensure common focus and effort.

There are many other barriers that can occur within organizations. For example, can staff gain efficient access to financial and other resources? Are there simple processes in place, based on commonly understood standards of performance, to obtain investments, hire staff, engage consultants, and build facilities? If not, the extra time and effort needed slow down the initiative.

Sometimes the barriers can be large and challenge the organization's image and existing structure. For example, it is increasingly important to obtain the best partners and resources from around the world. We are in a world of global innovation.[2] Many jobs can no longer be done cost-competitively in the United States. This can mean opening a new facility overseas, which can threaten existing employees who then drag their feet.

Obviously there are an unlimited number of issues that can prevent alignment. The key is to identify them quickly and then, when necessary, engage the extended enterprise to help remove them. We repeatedly discover that most of the self-protective "barriers" people believe are stopping them in an enterprise are not real. They are "red herrings." One common red herring is the belief that company-wide procedures prevent specific actions, such as being able to hire new staff quickly or form a new organizational structure. Another example, the perceived inability to get information from another part of the organization, is mostly illusory. Unfortunately, if your experience is like ours, one of the hardest things in the world is for a junior staff member to visit and talk to staff in another division without authorization. You might think that nothing would stop someone from talking to others, but if you remember back to the early stages of your career, you will realize that this simple act is a thousand times harder than it should be. It actually requires active support and enterprise-wide approval from management. We have been in many meetings at leading companies when someone asked one of their team members, "Have you talked with Jane in the other divison?" Invariably the answer is "No."

Removing these psychological barriers to gathering information takes time, but we have a simple tip. Write down your NABC value proposition and show it to your supervisor, pointing out where you need additional information and guidance from, say, an executive across the enterprise. First, this shows you have done your homework and you are not going to waste anyone's time. Second, it will reassure your supervisor that you are not doing an end run but rather that you have a legitimate business reason for talking to the other person.

We recommend that you write down all the barriers you see in front of you and then present them as part of your initiative at your Watering Hole. We have found that most staff across an enterprise, once the barrier is made clear, will step up to help smooth the way. Never assume something is not possible; do your homework and ask. Next we will discuss several other issues that aid in achieving alignment.

It's About Superior Results

Many enterprises introduce new ideas by talking about the need to change the organization's culture. We have worked with many hundreds of individuals, and so far no one has volunteered to have their culture changed. When management talks about changing someone's culture, it feels insulting and catalyzes resistance. Culture change is a by-product of developing new skills and demonstrating better results— it is not the starting focus of the Five Disciplines of Innovation. We *never* talk about changing culture.

The objective is to deliver superior results. As our Sarnoff example illustrated, the Five Disciples of Innovation do not initially have to involve the enterprise's management structure, political battles over titles and turf, or significant resources. The only thing that is required in the beginning is a modest commitment of time. But even here, all organizations have countless meetings. Many, if not most, of these meetings have incomplete agendas, they are not focused on creating customer value, the employees do not share common tools and language for quantifying customer value, and people do not come fully prepared. Over time, as people learn the Five Disciplines, there will be efficiency savings, because those meetings will become more productive. In addition, it does not take significant resources to create a regularly scheduled Watering Hole to focus on creating new value for your customers. Some money for pizza and soft drinks plus a few enthusiastic colleagues put you in business.

Consider another simple example: understanding customers' needs, which is the starting point for developing new customer value. Even if you don't have a budget to put people on planes to visit customers, you still have a telephone and e-mail, which allow you to interact quickly and efficiently to get needed information. These ready resources are generally not used aggressively enough.

The goal is steady, quantifiable progress toward implementation of the Five Disciplines of Innovation. Within your current team, you have all the resources needed to begin.

First Steps

The best way to start is to find a buddy. At Sarnoff, progress would have been extraordinarily more difficult if Winarsky had not been a partner. You will discover that having someone to discuss and adapt the Five Disciplines of Innovation, so they fit your team's context, is essential.

Second, as part of your work with your customers, find ways to evaluate the value you are currently providing them—using, for example, Value Factor Analysis. It is always illuminating and, frankly, often depressing to learn how far you have to go to do a superb job for your customers. Because I am CEO of SRI, for example, my key customers include all SRI employees. I need to continuously know if the organization is satisfying their needs. Through staff surveys, we have learned that we must continuously improve our performance. Some of the areas of improvement needed are hardy perennials: we need to redouble our efforts to communicate our vision, strategy, and goals; we need to demonstrate how everyone fits in and is valued; and we need to involve the staff more in important decisions. Improvement never stops.

Thoroughly engage your team as you introduce the Five Disciplines of Innovation. You should first get agreement on the need to create greater customer value and then to achieve this by using innovation best practices. As you engage your team in the use of innovation best practices, ask at every step, "What is the alternative? Are there better ways for us to engage our customers and understand their needs? Are there better words and tools to capture customer value? Are there value-creation processes that are more efficient?" Continuously identifying and using superior ideas is the essence of innovation best practices.

As you move forward, look for easy successes and engage the early adopters. Even with positive new ideas, only roughly 20 percent of the people will quickly engage, 70 to 75 percent will watch for proof to see if the idea works, and 5 to 10 percent will never join. You want to immediately engage the 20 percent who get it, help them create greater success, and then leverage their successes to engage the wider enterprise.

As you work with these ideas, you may discover that certain words or concepts need to be modified to fit your environment. We mentioned earlier the BBC, who liked the idea of NABC but added a small "a" out in front to represent the audience—aNABC. When people in the BBC read this, they say, "We must understand the audience needs (aN) to create our approach (A) for programming that will create the greatest public benefits (B) for the cost when compared with our competition (C) and all the alternatives people have today and in the future."

In addition, you will need to translate the examples in this book into ones that are appropriate for your enterprise. Examples from the automobile industry may not seem applicable to a finance department, a nonprofit, or the government, even when the underlying principles are the same in every setting. You and your colleagues will have to create examples that resonate with your team.

This book is about embracing innovation best practices throughout your team and then your enterprise. In addition to the ideas shared here, there are many best practices that apply to different functions within an enterprise. Examples include Design for Six Sigma, ISO 9001, and standardized work. Best practices often refer to *current* best practices, and of course you should copy best practices from any industry or enterprise that has them. Toyota, for example, developed the idea of just-in-time manufacturing by noting that produce in grocery stores needed to be delivered exactly on time or it would spoil.[3] Toyota then asked, "Why can't we do that with automobile parts? It would cut down on waste and reduce costs." That was the beginning of a revolution in automobile manufacturing. Over time, you should strive to develop innovation best practices that are unique to your enterprise and that go beyond anything others are doing.

Muda

There is a wonderful Japanese word for non-value-adding activities—*muda*.[4] Muda is the flip side of customer value. It is all the things that create waste and interfere with the creation of customer value. To have the biggest impact, it is not enough just to create the highest customer value; we must also eliminate the activities and unnecessary costs that get in the way. Everyone has, for example, been frustrated by going to endless meetings that didn't result in any action to help move the team forward. That is muda.

The wonderful thing about the word *muda* is that it sounds exactly like what it is. When your team members get to the point where they can look at one another and say, "That was muda" and everyone laughs, you are making progress. Muda also acknowledges that every time we create a new project we must then clean up all the parts that are no longer relevant in order to move forward. It's like baking a cake. The cake may provide enormous customer value, but when you are done you have to clean up the dishes before creating the next one. Obviously, one of your keys for helping eliminate muda is to be constantly talking about whether a given activity is important or interesting. Occasionally, you will discover that some activities are not even interesting. They provide no value at all for the enterprise.

Possible Mistakes

While introducing the Five Disciplines of Innovation, as with any initiative that involves the way people work, there are common mistakes that make progress harder. Although you are introducing processes to help your team be more successful, it is a mistake to think you won't be challenged. Employees may fear that the proposed change will negatively impact their jobs. At every step you will be tested and even misinterpreted. This happens because at first people are unsure of your full intentions and play out their worries. You must involve them and listen to the feedback you get. By continuously reframing their comments to confirm your understanding and then involving them, you will be able to move forward.

A second mistake is to make too big a deal about innovation best practices before you can show positive results. As your first act, don't buy everyone a T-shirt saying, "I'm a Champion" or "I Create Superior Customer Value." These unnecessarily flamboyant gestures will just provoke the ire of your colleagues. And no matter how effective you and your program are, results take time to achieve. Promising more than can be delivered is guaranteed to make your job harder.

A third mistake is that parts of the Five Disciplines of Innovation are either misapplied or applied so tepidly by your team that there is no progress. For some, this will prove that these value-creation ideas don't work. Be a champion to help your team fully organize the new skills and succeed. Preserve and create the customer value that will make your team the most valuable unit in your enterprise.

In the beginning, most members of your larger organization will likely disregard you. Then, when they see you having success, they may ask you to modify your practices to better fit their agenda. We had a dramatic example of this when we first started at Sarnoff. A vice president told us that there was no future in what we were doing and that he was going to focus on other parts of the business. We said, "That's okay. Thanks for telling us." Over time, our group began growing twice as fast as the rest of the enterprise. The same VP then showed up in our office for the first time, sat down, and said, "I have no idea how you folks are doing it. You keep on talking about important problems, compelling customer value, and all that other stuff. But if you would just lower your standards, there's a lot of good, average work out there that we could get to help grow the business faster." He didn't get what we were about. We politely but firmly asked him to never talk to any member of our team. Shortly after this meeting he became frustrated with his lack of success and left.

Show Progress—Others Will Notice

You can start your team on the path to creating greater customer value without permission from anyone else in your organization. As you develop your innovation best practices, there will be plenty of opportunities to influence your colleagues across the enterprise. If all your presentations are compelling, quantitative value propositions, they will be noticed by all. The opportunity to inspire your colleagues is actually fairly easy, because in most enterprises it is the rare presentation that starts out with *quantitative customer needs* and ends with a discussion of the *competition and alternatives*.

When people begin to notice your progress, it is then time to ask for additional resources. For example, although the innovation processes we describe here do not require huge investments, it is helpful to have a small pot of money to buy pizza, send team members to visit customers, buy market reports, go to conferences to study the competition, and hire outside Jungle Guides. Incredibly, in many big companies one of the first things they do when they have a financial downturn is to cut the travel budget. Okay, but the travel budget for visiting customers should not be cut.

The most valuable enterprise resource you have is your colleagues. Only by tapping into the full genius of your extended team can you respond fast enough in the exponential economy to thrive. Once they see that you are serious, most will help with your innovation.

Our perspective is inherently optimistic. We believe that this optimism is good business. People want to engage in things that matter to give meaning to their work; they want to do a good job because they desire being valued and appreciated; they want to develop new skills to become more valuable, regardless of their position in the enterprise; and they want to have healthy relationships with others because they know it is both more productive and fun. Your biggest motivating influence is to tap into these basic human needs—the desire to do an excellent job and make a contribution. You will be tested to see if you believe this. But deep down, most of your team members will hope that you do, because that is the direction they want to go in too. Help them.

the INNOVATION enterprise:

continuous value creation (cvc) throughout

"WHENEVER SOMETHING—ANYTHING—IS TO BE
PRODUCED, THERE MUST BE A SYSTEMATIZED
METHOD OF PRODUCING IT."[1]

Toyota employee handbook

"EVERY CEO WILL AT LEAST GIVE LIP SERVICE TO
THE IDEA THAT THE WORLD IS MOVING FASTER AND THAT
WE NEED TO DO A BETTER JOB AT INNOVATION.
BUT IF YOU GO INTO AN ORGANIZATION AND ASK PEOPLE
TO DESCRIBE THEIR INNOVATION SYSTEM,
YOU GET BLANK LOOKS. THEY HAVE NONE."[2]

Gary Hamel, Harvard University

Quality Is a Given

When you are in the role of a customer, you certainly expect to put
your money down on quality products. Many of us, for example, go
to various consumer reports and review websites to select our next
automobile—hoisting quality up to first place. We read news articles,

ask friends, "How do you like your new car?" and seek out other indices of automotive quality, such as J.D. Power ratings and *Consumer Reports* repair records.

We expect quality and are shocked when our car doesn't start, pen doesn't write, PDA goes blank, or TV goes on the blink. It wasn't always like this. Before product companies adopted the principles of Total Quality Management (TQM), many goods didn't pass the quality test. As a result of worldwide acceptance of TQM, everything from computers to automobiles now achieves impressive levels of quality.

When it comes to the production of goods, quality is king. Manufacturing quality is now the minimum expected in order to enter the competitive arena. But for knowledge workers, your customers demand more than just quality. Customer value, as we noted in Chapter 4, "Creating Customer Value: Your Only Job," ranges all the way from product features to deeper needs for identity and meaning. Customers increasingly demand products and services that not only work but also provide superior experiences and enhanced prestige, and fulfill other needs as well. The quality movement in manufacturing needs to be revisited to extend TQM.

An Inspiration

They couldn't believe what I was saying.* At an SRI executive staff meeting I mentioned that we were going to visit an automobile manufacturing plant. "Why in the world would we do that?" they asked. "We don't make automobiles; we conduct research and create new technological innovations and solutions." I agreed, but I wanted everyone to see an example of the power that comes from having a process for innovation in manufacturing.

In 1982, the GM manufacturing plant in Fremont, California, across the bay from Silicon Valley, was one of the poorest performing plants in the country. Its products had many defects, and the employees and management were not working well together. Just before it was closed,

---

* This section is in the voice of coauthor Curtis R. Carlson.

it had almost 800 union grievances pending and an absentee rate that reached 20 percent.[3]

In 1987, Toyota and GM agreed to form a 50/50 joint venture (JV) and reopen the plant under Toyota's management. The first CEO of the plant was Tatsuro Toyoda. The JV was called New United Motor Manufacturing Incorporated, or NUMMI. The NUMMI team renegotiated the existing union agreement to accommodate Toyota's broad job classifications and flexible work rules. Of the original employees, roughly 50 percent were hired back as the core of a reduced workforce.[4]

The NUMMI JV was formed at a time when there was little understanding of the Toyota manufacturing system. One senior GM manager said, "Initially, we thought there was nothing to learn from our partner. When we first went to Japan, we thought our partners wanted a JV so they could learn from us. We were shocked at what we saw on that first visit. We were amazed that they were even close to us, let alone much better. We realized our production capabilities were nothing. Our partner was doing many things we couldn't do."[5]

Several years after the NUMMI plant reopened, it had the highest quality and productivity in the GM U.S. organization. Total worker hours per vehicle averaged 21 at NUMMI versus 43 at the same plant in 1978. A Toyota plant in Takaoka, Japan, averaged 18, and the comparable GM plant in Framingham, Massachusetts, averaged 41. Absenteeism was down to 3 percent and union grievances were about 15.

Toyota Standardized Work

How did Toyota and GM achieve such a remarkable turnaround at the NUMMI plant? Like the success of Engelbart's team in inventing the computer mouse and the foundations of personal computing, an achievement of this magnitude does not happen by accident. Later we learned about some of the reasons for Toyota's success. Its philosophy is nicely summarized by this quote from the Toyota handbook given to all employees: "Whenever something—anything—is to be produced, there must be a systematized method of producing it. Whether or not the people who do the actual production fully understand those rules, that system has a deciding effect on product quality, cost, safety, and all essential determinants of success or failure."

These are all important words, but the one that stands out is *anything*. That includes innovation. Toyota then goes on to tell all its employees:

> Toyota's mission is to provide to society:
> - the highest-quality automobile
> - at the lowest possible cost
> - in a timely manner with the shortest possible lead time
> - while paying the utmost respect to the humanity of workers
> - through the use of Toyota Standardized Work
> - and with the elimination of all muda—non-value-adding activities.

This is the best mission statement we have seen. The first three bullets define value to the customer, based fundamentally on quality and speed of production. The next three bullets say how it will be achieved.

At the core of Toyota Standardized Work is the idea of Total Quality Management (TQM). In a conversation we had with Tatsuhiko Yoshimura, former vice president of quality for Toyota, he emphasized that TQM is the foundation for all activities at Toyota. He drew a picture for us with TQM as the base that firmly holds up all programs and initiatives; without such a strong base, major strategic initiatives would tip over and fail.

Ford, Deming, and Ohno

The last hundred years have seen two revolutions in the processes used for manufacturing innovation. The first was when Ford pioneered *low-cost* assembly-line manufacturing with the Model T in 1908.[6] The second was the introduction of Total Quality Management (TQM) to manufacturing, by Deming, Ohno, and others in the 1950s.[7] Ohno's lean manufacturing system went beyond Deming and was the basis for the Toyota Production System. These innovations by Deming and

Ohno added *high quality* to low-cost manufactured products. Manufacturing companies must focus on quality, because it is the most important element of a product's value. If your car is always breaking down, no matter how luxurious it is you will be very unhappy. Today all manufacturing companies use some version of TQM to both continuously increase quality and reduce costs.

NUMMI demonstrated the power and efficiency that come from having a disciplined quality process. The NUMMI success was based on best practices focused primarily on quality, which is the core attribute that a manufacturing facility must deliver. In the exponential economy, manufacturing quality represents the "table stakes" required for entry into the game. In many areas, manufacturing is becoming almost a commodity that can be purchased. *Successful enterprises must go beyond quality to the continuous creation of new customer value in the shortest time and for the minimum investment as the overarching organizing principle.*

We are now firmly into the global knowledge age—the exponential economy. There is currently no equivalent of TQM for knowledge workers. And for knowledge workers, the right metric is customer value, which explicitly encompasses all of a product's attributes, including quality, service, experience, and emotions. It is also a metric that applies to every function of the enterprise. By analogy to TQM, we believe that the proper foundation for all knowledge enterprises is Continuous Value Creation (CVC), where every function of the enterprise is dedicated to creating the highest customer value in the shortest possible time.

CVC Is Overarching

Continuous Value Creation (CVC) develops an enterprise-wide culture that unrelentingly focuses on the rapid creation of new customer value. CVC provides for both continuous improvement and the development of large, transformational innovations. Because the creation of customer value is the primary function of an enterprise, CVC includes the overarching family of best practices into which all others fit. TQM best practices, such as Design for Six Sigma and ISO 9001, are more specific and apply to selected parts of an enterprise, whereas CVC is at the core of innovation best practices throughout the enterprise.

TQM and CVC have complementary properties:

TQM	CVC
Objective	
Consistent incremental quality improvements	Continual customer value improvements through both incremental and breakthrough innovations
Metrics	
Statistical measures of quality (i.e., one element of value quantified)	Customer value created per time per $ invested (i.e., all elements of value)
Focus	
Quality, efficiency, and cost	Increased customer value (benefits per costs), efficiency, and speed of innovation
Leverage	
Continuous process improvement through rapid iteration and compounding	Continuous process improvement to deliver new customer value through rapid iteration and exponential compounding

Bottom Up and Top Down

Continuous Value Creation allows for both top-down and bottom-up initiatives. Organizations must do both. Bottom-up is essential because your employees are the most knowledgeable about customer needs, changes in businesses models, and revolutions in technology. Top-down is essential because major new initiatives are also required, such as GM starting a fuel-cell program to replace the internal combustion engine.

An early American contribution to management was the development of top-down strategies, driven through the whole organization. One extreme approach was represented by Henry Ford, who obsessively organized his company from the top down, with employees as "cogs" in his low-cost, mass-production assembly line. He did not want his managers to make decisions. Their job was just to implement his rules and orders. If you did act independently, you were fired. He was fixated on achieving low cost and he thought this was the best way to achieve it. He only offered cars in black because one color simplified production and black paint covered best. While he did produce the revolutionary Model T, he also initially had a 40 percent per month worker turnover rate from bored, burned-out employees.[8] People didn't like being treated like robots. They were not fully engaged and, undoubtedly, quality suffered.

Top-down organizations can be responsive, but they tend to be based on many specific rules to produce the required results. Ultimately, they fail because the organization's collective intelligence is grossly inadequate to keep up with the external pace of innovation. When GM offered cars in different colors plus other innovations, Ford almost went bankrupt.

Total quality management (TQM), in contrast, begins by focusing on the quality of work produced. Since it is aimed at continuous improvement throughout the enterprise, it requires that all workers be involved. In the Japanese application of TQM, top management provides leadership through the articulation of mission statements and a total commitment to TQM as the basis for all activities. Typically, in a Japanese TQM enterprise, you hear phrases like "to bring delight to our customers" or "the imperative to continually contribute to the development of society while focusing on corporate responsibility."[9] Responsibility is distributed throughout the enterprise, but new initiatives must be extensively socialized before they become accepted.

The Japanese have a wonderful term for this form of socialization, *nemawashi,* which literally means "going around the roots." The concept comes from growing bonsai trees. With bonsai, you have to ensure effective root binding for the plants to thrive. More time is spent on developing the roots than on pruning the top of the tree. In management terms, this means spending time on getting people engaged and bringing them to a consensus.[10] The concern in these enterprises is whether the enterprise will respond quickly enough to seize the rapidly developing opportunities of the exponential economy.

An enterprise based on CVC works to harness the power of *both* top-down and bottom-up. In CVC organizations, the creation of customer value is the core around which all other activities are built. Other innovation best practices, such as TQM, are included throughout the organization as appropriate.

With CVC there is room for strategic initiatives from the top as well as new ideas percolating up from the bottom. Watering Holes provide a common organizational structure throughout the enterprise to bring forth both types of initiatives. For example, Watering Holes can be run at the lower levels of the organization, focused on improving today's products and services. Alternatively, the upper levels of the organization can form Watering Holes around new strategic initiatives, based on transformational technologies or business models, and place them at the proper place in the organization. The central driving feature for all, however, is the focus on delivering the highest *customer value* in the *shortest possible time* with the *minimum resources* using *innovation best practices*. New initiatives—whether from the top, the bottom, or the middle—all have to pass this same litmus test.

SRI's First Steps Toward CVC

Moving to SRI's headquarters in Silicon Valley to become the CEO of one of the premier technical enterprises in the world, was a dream come true. There was just one problem: The business model that had been so successful for SRI over the previous fifty years was not keeping up. The vitality and speed of the exponential economy required SRI's teams and partners to work together in new, more productive ways.

It is always a mistake to walk into an enterprise and assume you understand it. I certainly didn't understand what was important to SRI's employees and I needed to find out. When I arrived, I prepared a questionnaire with Alice Resnick, our communications vice president, which we gave to all employees. It asked everyone:

- What are our vision, strategy, and key values?
- What is important about SRI and our work?
- What are our strengths and weaknesses?

The hundreds of pages of responses I received were impressive. Clearly the staff cared deeply about SRI. But there was little agreement about our vision or strategy. Fortunately, there was almost universal agreement on basic values. "We are customer-centric." "We do important work and make a positive impact on the world." "We work with outstanding professional colleagues." "Respect and integrity are fundamental to everything we do." I summarized the comments and shared them with all the staff members to make sure I had correctly understood their feedback.

The senior staff and working groups throughout the organization began to explore SRI's external environment and how we needed to respond. We started by working to understand our customers' needs, our overall business environment and ecosystem, and our specific opportunities. We asked questions like "Are we in the right markets? Do we have the organizational structures to deliver on our opportunities? Do we have the people to succeed, and do we have the appropriate support infrastructure?" Out of this work, we began to develop a new vision for where we wanted to go and the work that was required to get there. The senior staff was asked to take on new roles to move us toward the new vision. Additional support teams were formed to create the missing initiatives, organizations, and infrastructure and to modify those that were no longer appropriate.

Speed was essential, but it was also critical that we not get too far in front of the entire organization: we needed the staff's input and buy-in. Enterprise-wide talks, focus groups, team meetings, and e-mail were continually used to involve others and solicit feedback.

The DNA of change (*d*esire, *n*ew vision, and *a*ction plan) not only applies to individuals, it plays out in organizations too. In the beginning at SRI, there was little agreement about the need to make any change. People would say, "Our team is fine. Just get that other group to shape up." But, slowly, the enterprise began to understand the unique issues we confronted in the exponential economy and why we needed to change. After a while, people stopped talking about the need and began talking about SRI's new vision and where it should go. Eventually, the discussion moved from the new vision to the action plan, using the steps described below.

We started with two overarching principles: The first was a commitment to creating and delivering the highest customer value in the shortest time. The marketplace demands this. The second was to achieve this goal through the use of innovation best practices. Our test was to always determine whether our practices would move the enterprise in the direction of creating greater customer value. If they didn't work out, we tried others. The process was one of *continuous* improvement of our innovation best practices through discover and create, apply and measure, and learn and iterate.

These two overarching principles are the most significant acts of commitment toward achieving CVC. We see hesitancy and procrastination about making these commitments all the time in other enterprises, perhaps because the ramifications of these decisions are unfamiliar. But we sometimes paraphrase a well-known quote by asking, "If not this, what? If not now, when? If not you, who?"[11] We often get blank looks. We conclude that these commitments take great courage by the CEO and the senior management team. But what is the alternative to being explicit about creating the highest customer value using innovation best practices?

SRI Card

One of our first deliverables in our work at aligning SRI around customer value and innovation best practices was the SRI Card, which we developed with our staff. The card would articulate SRI's vision and strategy along with our values and key innovation practices. Working groups began developing the card. Periodically, versions of the card were placed on the internal SRI website to get a company-wide response. In the beginning, some of the e-mail responses were argumentative, suspicious, and even antagonistic. But all the responses were helpful. They indicated that we still did not have the right messages and we needed to do more work.

Iterations continued. The goal was *not* to produce a card in record time, but rather to involve all the employees in a joint vision that would work. The SRI Card was a conversation starter and vision clarifier. Finally, e-mails started coming back saying, "That's who we are." "This fits." At this point we published the card shown in Figure 16.1.

Figure 16.1: The SRI Card given to every SRI employee, which includes key elements of the Five Disciplines of Innovation. *BusinessWeek* dubbed SRI the "Soul of Silicon Valley" because of its 60 years of pioneering contributions.

After finishing the SRI Card, we continued our working groups and focused them on key innovation issues, such as the roles of champions, how government work differs from commercial work, and the appropriate quantitative goals for SRI. We created new organizational structures, such as Watering Holes and a ventures-incubation group. We added new people to our staff, and a few, who didn't share our values, left. In the process, we created the SRI Discipline of Innovation workshop to make these innovation skills available to our staff, partners, and clients.

The SRI Card was an early visible sign of our commitment to alignment in creating the highest customer value using innovation best practices throughout the enterprise. But alignment in an enterprise, where everyone understands what customer value and innovation best practices mean and can act on them, is a long-term project.[12] First, it means that the entire enterprise is aligned around the Five Disciplines of Innovation and has a deep commitment to them. Second, it requires that all functions of the enterprise be designed to allow the creation and delivery of the highest customer value. As is often quoted, "All organizations are perfectly designed to get the results they get too well."[13] Just like the challenge at SRI, you have to align all the key elements in your enterprise, as listed below.

——{ KEY ELEMENTS OF ORGANIZATIONAL ALIGNMENT }——

- Shared vision, strategy, values, and goals
- Commitment to delivering the highest customer value
- Commitment to continuous improvement processes using innovation best practices: the Five Disciplines
- Creation of innovation organizational structures and processes with the appropriate staff
- Organizational transparency, including staff communication, knowledge, and intellectual property
- Shared recognition and rewards
- Commitment by the CEO, president, and senior management team

Our objective was to achieve CVC across SRI. In a CVC enterprise, all employees create and deliver new customer value. This expectation respects the importance of each person's activity. In addition, consideration is given to all the stakeholders to assure that they are in balance: customer value, company value, shareholder value, partner value, employee value, and public value.

Measures of success for a CVC enterprise include:

- New customer value created and speed of innovation
- Customer satisfaction and repeat business
- Growth, profitability, and return on investment
- Employee development, loyalty, and retention
- Community responsibility

Alignment means assuring that the enterprise is constructed to deliver the highest customer value. If organizational issues are in the way, they must be removed. Your goal in a CVC enterprise is to use innovation best practices to make the delivery of the highest customer value *inevitable.*

Early Adopters

Employing CVC throughout the enterprise is like any other initiative: not everyone will join on day one. Start by finding those who share your passion for the new vision, which is much like establishing a beachhead in the marketplace. Help them make the transition, achieve new levels of performance, and celebrate their successes. There is nothing that builds credibility more than having respected members of the organization embrace the new vision, master the new skills, and succeed. These early champions prove to the skeptics that CVC works in a way that talking cannot.

In addition, if you want the principles of customer value and innovation best practices to stick, you need to continuously reinforce them. As we have mentioned, you must say things ten times more often than you would ever believe, because not everyone listens, the words used are unfamiliar, and there is skepticism at first. It also takes time to understand and internalize new concepts. For example, understanding how to create a quantitative value proposition is neither easy nor quick. Since different people learn differently, we used diverse approaches to engage everyone, such as all-staff meetings, working groups, and training, while posting our results on the SRI Internet.

Your CVC Road Map

We are often asked, "How do I begin?" The following overall road map is one way to glimpse the entire movement to CVC for your enterprise. It gives specific steps you can use to bring CVC into focus. Of course, in your enterprise you will have to adapt these steps, but this overall road map will point the way for your CVC migration. As the elements of the enterprise become aligned around the Five Disciplines of Innovation, they mutually reinforce one another and progress will accelerate:

1. Deeply understand the exponential economy and its impact on your enterprise.
   - *Select your important customer and market opportunities.*
   - *Create a vision of where the enterprise must go to thrive.*

2. Create support groups to assure that the enterprise realizes the new vision.
   - Prepare for the journey.
     - *Take an inventory of where you are.*
     - *Begin to involve your team and get buy-in.*
   - Educate and enroll others in the vision.
     - *Use every effective vehicle possible to communicate the new vision and value-creation strategies.*
     - *Create an Elevator Pitch for your team and organization around the new vision.*
   - Launch your initial Watering Holes.
     - *Select a business theme, prepare champions, and invite key contributors.*
     - *Run the Watering Hole and educate participants about value creation while removing obstacles to collaboration and value creation.*
     - *Encourage others to develop value propositions and conduct Watering Holes.*
   - Create momentum through short-term wins.
     - *Recognize and reward people for creating new value and supporting the vision.*
     - *Find and coach other champions and keep others informed of the progress.*

3. Align the enterprise with the new vision and institutionalize innovation best practices.
   - Consolidate gains and produce still more change.
     - *Use credibility to change systems, organizational structures, and policies to fit the vision.*
     - *Educate and train champions and staff on value-creation principles and practices.*
   - Create support structures.

- *Keep the voice of customer value and continuous improvement alive.*
- *Document and share lessons learned and other innovation best practices.*
- *Align rewards and recognition programs with the vision.*

CVC, like all organizational approaches, must be continually reinforced. It is a journey that takes time. Look for opportunities to make headway, and come back at it again and again. We all think it will happen sooner than it does. It takes time for even a positive virus to infect the entire enterprise.

Jim Collins has a wonderful metaphor for organizations that move from "good to great":[14] An enterprise is like a huge flywheel. When you first push on it, it barely moves. But every little push builds momentum and eventually the wheel will spin. Great companies are the ones that align the enterprise and keep the "little pushes" going in the right direction. Failure comes from thinking that one or two big pushes will get the flywheel moving fast enough and keep it going.

We use another metaphor that helps us measure progress: "brick by brick." If your aim is to build a solid wall, you do that brick by brick. We constantly ask whether we are adding more bricks each week, each month, and each year. At first you don't see much progress, but eventually everyone can see the progress that has been made. And once you get going, you will start adding bigger bricks.

We are often asked, "What percentage of SRI International's staff fully understands these principles?"

We say, "About 10 percent understand deeply, 20 percent have a good working understanding, and another 30 percent are conversant with the ideas. The remaining 40 percent are familiar with some of the ideas, like NABC." It's like TQM that way; it took Toyota fifty years to get where they are. But you don't need 100 percent understanding of these ideas to make a significant impact and improve the success of your enterprise. You can never achieve 100 percent, but as the parts come together, they amplify one another and the impact increases. Steady progress brings accelerating rewards.

Tests of Leadership Commitment[15]

If you are a leader promoting CVC, there will be many tests of your commitment. As you begin talking about customer value, the process of innovation, and how to create and deliver superior customer value, others will test your leadership. If there is any disagreement within the upper leadership team, including the president and CEO, it is a huge red stop sign for the rest of the enterprise. People will think, "If they aren't on board, why should I pay attention?" The entire management team will be continuously tested. You will be in a meeting, for example, and someone will say, "Oh, come on, this is just another management trend du jour." Any hesitancy on your part to respond appropriately is a failed test of commitment. The president, CEO, and the entire management team must be champions for the Five Disciplines of Innovation.

Employees will watch you like a hawk to see if you waver or if you really are a champion for the new approach. Do you have enthusiasm? Can you articulate the Five Disciplines of Innovation? Can you engage them in conversation so they become convinced of its utility to *their* jobs? When you respond with clarity and enthusiasm, always taking account of the individual and team concerns, you can incrementally enroll everyone who is interested in higher-impact innovation through CVC.

Before others will join you they need to know you are going to continue to lead in the same direction for more than the typical half-life of most initiatives. If you change your mind every year, they will just wait you out. Recall all the catchphrases that have appeared and receded in the recent past:

- Management by Objectives
- Strategic Planning
- Management by Walking Around
- One-Minute Manager
- Zero-Based Budgeting

There is much that is right with these approaches, but each one is only part of the puzzle that makes up the Five Disciplines of Innovation. If you keep jumping between initiatives, others will see you as too fickle to invest in. What is needed is a comprehensive, overarching approach where everything you ask of others makes sense to them. Creating and delivering the highest customer value, continuous improvement, innovation best practices, and CVC provide that framework. For all initiatives, we ask three straightforward questions:

- Is this action moving us toward creating greater customer value faster?
- Is there a better alternative? If so, let's use it.
- Are there non-value-adding activities—muda—we can eliminate?

Employees will test your commitment by looking for areas of contradiction. As part of your CVC inventory, check for problem areas. Then, if you find any, put processes in place to address them. Common areas of misalignment include rewards—which must support productive teamwork—and senior management—who must all exhibit the values of CVC.

Steady Progress Wins

Can you do this? Yes. The key is steady, solid progress. As one part of the organization aligns with another, you will experience positive, compounding improvements. If NUMMI can rise from the dead, others can too. After all, your customers are waiting.

CHAPTER 17

INNOVATION'S
five DISCIPLINES:
a foundation for national
competitiveness in a
world of abundance

"WE [THE UNITED STATES] ARE NOT COMPETITIVE."[1]

John Chambers, president and CEO of Cisco Systems

The Silicon Valley Model

I moved to Silicon Valley from Princeton, New Jersey.* Princeton is an enchanting town full of accomplished people, but if you are interested in innovation or technology, Silicon Valley is the center of the world. In ratings of entrepreneurial excellence in the United States, there is Silicon Valley and then, down the list, there are Boston's Route 128, Research Triangle Park in North Carolina, San Diego, and Austin. Silicon Valley, for example, has Sand Hill Road, which is only a few miles long, but it contains the addresses of many of the country's leading venture-capital firms. Each year, $8 to $12 billion are invested from Sand Hill Road and the surrounding community—roughly 33 percent

* This section is in the voice of coauthor Curtis R. Carlson.

of all U.S. venture capital.[2] Boston and all of New England, for comparison, provide around 11 percent. All of Texas provides just 3.5 percent.[3]

Silicon Valley is unlike any other place in the world in other ways. What everyone notices first is the almost perfect weather, the spectacular countryside, and the terrific variety of fresh, ethnic foods. But Silicon Valley is most distinguished by being a meritocracy: who your parents are, who you know, and where you went to school are not what matter most. And although issues of race, gender, religion, and sexual orientation never disappear completely, they are less significant in Silicon Valley. What ultimately matters are your skills, values, and commitment to the task. As one example, more than 30 percent of Silicon Valley CEOs are émigrés from China and India.

Silicon Valley is dedicated to creating and delivering new customer value. It is different from Washington, D.C., which is about politics; New York City, which is about finance; or Hollywood, which is about entertainment. Of course these locations innovate and provide great value for their customers too. But in Silicon Valley you make your mark primarily by creating new high-impact products and services that are successful in the marketplace. It looks, feels, smells, and tastes different. Silicon Valley is liberating and empowering. It is a world of abundance.

Visitors from other U.S. states and countries come to Silicon Valley to learn how to create their own Silicon Valleys. Whether you're from Bangalore, Shanghai, or Dublin, Silicon Valley is a model for innovation and productivity. But increasingly, Silicon Valley needs to learn from others. In the exponential economy, if you are not improving, you are losing ground.

Everyone wants to live in a prosperous country and have a high quality of life, a sense of security, and opportunities for future generations. Prosperity is the result of competitiveness and productivity, which are based increasingly on innovation. Huge wealth from natural resources is usually not sufficient to bring about widespread prosperity in a country. Take Saudi Arabia, most of the rest of the Arab world, and Russia. They are swimming in oil and yet they have marginally successful economies.[4] Great wealth flows from success in the exponential economy by creating higher-value products and services for customers. Consequently, increasing the innovative capability of regions around the world is a central issue for most governments. No one is immune from the forces of innovation in the exponential economy, not even the United States. Unfortunately there are many reasons to believe that the relative competitiveness of the nation is in decline and that remedial actions are required immediately.

Need to Improve Innovation

Cisco Systems is one of the world's largest companies, with sales of more than $15 billion per year. It builds hardware and software that form the backbone of the Internet. Cisco is the quintessential global company in the exponential economy. It works with the best partners around the world and seeks out the smartest, most talented, and most highly motivated employees. When John Chambers, the CEO of Cisco, says, "We [the United States] are no longer competitive," it is time to pay attention. Chambers sees firsthand what is happening in India, China, and elsewhere and compares it with what's happening in the United States. He's obviously not impressed with the United States. The most recent quantitative index of competitiveness shows that in 2005, the United States will have had the largest drop in competitiveness ranking of any country in the world, moving from first in 1995 to sixth in 2005.[5] And in a survey of U.S. business executives, they were only "slightly positive" about the innovation climate in the United States, but significantly more enthusiastic about innovation prospects globally.[6] These executives are reflecting the reality that the rest of the world is picking up the pace. Consider that as of 2005:

- Foreign-owned companies and foreign-born inventors account for nearly half of all U.S. patents.

- Sweden, Finland, Israel, Japan, and South Korea each spend more on R&D as a share of GDP than the United States.

- Only 6 of the world's 25 most competitive information-technology companies are based in the United States; 14 are based in Asia.

- China overtook the United States in 2003 as the top global recipient of foreign direct investment.

- Asia now spends as much on nanotechnology as the United States.[7]

These are some of the items that Chambers was talking about. They do not represent the entire U.S. economy but rather the sharp point of the spear—the part of our economy that thrives on innovation and that will be responsible for prosperity, opportunity, quality of life, and security in the future.

The Innovation Workforce

Consider the U.S. educational system. Noted venture capitalist John Doerr calls education "the largest and most screwed-up part of the American economy."[8] While some might feel that Doerr is overstating the case, he is right that the U.S. education system is falling behind. The Program for International Assessment[9] found that the United States ranked twenty-fourth out of twenty-nine nations in math skills for fifteen-year-olds. Only 41 percent of *nonpoor* fourth-graders can read proficiently in this country.[10] And worse, only 15 percent of *poor* fourth-grade students can read proficiently. The gap among the most gifted students in the United States compared with other countries is even larger than for poor students.[11] And the celebrated pioneer of modern management, Peter Drucker, said, in 1997, "Thirty years from now the big university campuses will be relics. Universities won't survive. It's as large a change as when we first got the printed book. Do you realize that the cost of higher education has risen as fast as the cost of health care? Such totally uncontrollable expenditures, without any visible improvement in either the content or the quality of education, means that the system is rapidly becoming untenable. Higher education is in deep crisis. Already we are beginning to deliver more lectures and classes off campus via satellite or two-way video at a fraction of the cost. The college won't survive as a residential institution."[12] Again, you might feel these statements are extreme, but big changes are certainly under way.

There is no easy excuse for the poor educational performance in the United States. Consider California, which is the world's current exemplar for innovation and entrepreneurship. If it were a country, California would be the eighth-largest economy in the world, after the United States, Japan, Germany, the U.K., France, Italy, and China.[13] But K–12 education in California ranks fourth from the bottom of the 50 states.[14] For decades, California has made up for this abysmal educational performance by attracting brilliant students from around the world to its great research universities, such as Stanford, UC Berkeley, UC Los Angeles, UC San Diego, Caltech, and UC San Francisco. These are among the top twenty universities in the world.[15] But according to the president of Stanford, John Hennessy, "In the past, three out of four of our Chinese students would stay in the United States But this is no longer true, because students increasingly have great opportunities back home."[16]

Nations around the world are turning out large numbers of well-educated graduates qualified for work in the exponential economy. China will produce about 3.3 million college graduates in 2005, India about 3.1 million English-speaking graduates, and the United States about 1.3 million. In science and engineering, China will graduate more than 600,000 students (18 percent of China's total graduates), India 350,000 (11 percent), and America only about 70,000 (5 percent). To make things worse, the number of U.S. science and engineering Ph.D. graduates has started declining. Over the past ten years, the number of computer-science Ph.D. graduates has declined by more than 25 percent.[17] At the same time, China has a goal of producing a million science, math, and engineering graduates a year by 2010.[18] These skills are essential in the exponential economy.

Will the United States be the dominant technology nation thirty years from now? It is absurd to think that U.S. graduates are ten times smarter and more productive than their Chinese counterparts. Furthermore, in the exponential economy China and India do not have to start their research programs at the bottom and slowly work up. China and India are teaming with the best partners around the world to develop the next waves of technology now. In some areas, such as aspects of nanotechnology, China is reaching parity.[19] India has achieved this in many areas of information technology.

Unfortunately, on top of these striking numbers, the United States does not have a culture of enthusiastic support for education. Imagine U.S. students studying for nine hours a day, six days a week, as they do in China. Chinese elementary and secondary students do not watch television on school nights—they study.

You might be wondering if the Chinese and Indian educational approaches are rote learning systems that do not produce creative people. Maybe, but that is not what we see. Or you might note that the percentages of Chinese and Indian engineers and scientists per capita are similar to those in the United States and that degrees have less worth in China and India and thus there is no real issue. There are two things wrong with that argument: First, they are still doing better on a percentage basis than the United States. Second, and more important, when you are producing a million graduates a year versus seventy thousand, eventually the center of mass of innovation will shift to where the talent resides. Consider that for decades the United States always won the Olympic gold medal in basketball. At the time, most of the best players lived in the United States, and many thought that the United States would be invincible forever. But in 2004, the United States came in third, behind Argentina and Italy. The same thing will happen with America's ability to innovate, absent actions to fix the situation.

The U.S. ace in the hole has always been its ability to attract the best and brightest from around the world. Currently, 59 percent of the U.S. science and engineering Ph.D. students are foreign nationals.[20] And, as the quality of life and opportunities increase in their home countries, they will increasingly go home. In addition it has become much harder for the best students to come to America after the 9/11 terrorist attack because of security concerns. H1-B visas, which allow highly qualified foreign workers to remain in the United States for as long as six years, have dropped from 195,000 to 65,000 a year.[21] As Geoffrey Colvin pointed out in *Fortune* magazine, "If Albert Einstein wanted to move in today but had no U.S. relatives he'd have to get in line behind thousands of poorly educated manual laborers who did."[22] How can this be in the interest of the United States? Who designed this muddle? Osama bin Laden is undoubtedly pleased to have further reduced America's competitiveness, but the rest of us cannot be.

A rational system would act as a magnet to attract even more of the best talent from around the world. About a million people move to the United States each year,[23] and roughly half of this total is illegal entrants, who are poor and relatively uneducated.[24] No other country is remotely this generous. It is folly not to attract a proportional number of well-educated people who can rapidly create the financial basis to absorb these undereducated people and provide them with the education and, eventually, the good jobs they deserve.

The United States has a host of other issues that must be addressed for it to thrive in the exponential economy. For example, one reason the United States has created great research universities is the extensive government support for both basic and applied research. But this advantage is going away. For thirty years federal funding of physical sciences research as a percent of GDP has been declining.[25] Indeed, basic sources of research support, such as the National Science Foundation, have reduced, rather than increased, funding.[26]

At every turn the U.S. government currently seems intent on making U.S. innovation and competitiveness worse, not better. New rules and regulations are piled on older ones with seeming obliviousness as to their impact on innovation, competitiveness, and job growth. Each new regulation, tax increase, and OSHA (Occupational Safety and Health Administration) requirement may be well intended, may provide some desired function, and may not be a showstopper, but they all represent costs that take their toll. In a rational system, the benefits of these additional rules and regulations would be weighed against both their direct and indirect costs in terms of future job growth.

Consider, as one example, the recent legislative response to the post-bubble acts of a few large-company executives at Enron and World-Com. This legislation is called Sarbanes-Oxley—or, for short, SOX. SOX mandates a family of additional auditing provisions that must be performed by all public companies.[27] Some of the provisions are useful reforms, but others are expensive and provide no significant value for the public, the company, or the shareholders. The hundreds of billions of dollars lost because of SOX is money that is not used for innovation and job growth.

SOX applies equally to both big and small companies, which creates special problems for small companies. It has been estimated that in the United States 80 percent of new jobs come from small companies.[28] To understand the special problem caused by SOX, imagine you have a new company that must go public to raise needed capital. Previously, you were able to go public when your company was close to profitability with revenues of $25 to $50 million a year. Now, the day after you go public, you are hit with a bill for about $1 million in SOX compliance costs.[29] For a small company, that is the difference between being profitable or not. It makes creating a new company riskier and more expensive, because the company must now be larger to absorb that additional cost. The unintended consequence is that fewer companies will be formed. SOX is a silent job killer. A smart U.S. policy concerned with innovation and competitiveness as the source of job growth would be much more judicious about legislation such as SOX. Big and small companies would be treated differently. As we have described, only a fraction of new innovative companies eventually survive. They should be treated like the easily damaged, delicate flowers struggling for life that they are.[30]

Unions, for example, have not yet adapted to the realities of the exponential economy. They must become partners with management to create increased customer value, not antagonists fighting every new initiative. The NUMMI example in chapter 16 shows it is possible. The rapid decline in union membership in the commercial sector over the past few decades seems like an inevitable response to their strategies, which evolved to address the management inhumanities of the Industrial Age and the likes of Henry Ford.

The United States is not alone in confronting these challenges. Consider Germany, where unemployment is over 11 percent and job growth has been neutral to negative for over a decade.[31] The reasons for this are not surprising. When you learn about German labor law, you see that it is enough to scare even the most intrepid entrepreneurs out of the country. For example, after six months an employee in Germany has a special layer of protection. The employee can be fired only if a small group of employees elected by their fellow employees agree with management's decision.[32] Management can always go to court, but that is a slow, costly process. Imagine what this means to a new start-up company, where quick decisions are critical to survival. Some entrepreneurs can work around this system, but it makes them more cautious about hiring. It is another straw on the camel's back to slow down the innovation engine.

German labor law is just one example. In Germany, taxes are high, and because of a lack of venture capital, entrepreneurs must often get their capital from banks—organizations not exactly celebrated for their risk tolerance. And culturally, both success and failure are socially problematic. Great success is seen as a sign of arrogance. Failure can mean the end of your career. Optimism about the future is decidedly muted in Germany. No one should be surprised that there are so few start-ups and so little job growth. Germans live in a world of scarcity.

But other European countries are rising to the challenge. Consider Ireland. Only twenty years ago, Ireland was at the bottom of the European economic ladder. As Thomas Friedman points out, it is remarkable that the European country with the second-highest per capita income is the "land of poets and balladeers."[33] (The first, Liechtenstein, is an anomaly, since it has a tiny population of 30,000 people.) Unemployment in Ireland is only 5.2 percent.

Ireland's progress is not a fluke. The Irish have embraced a family of innovation best practices to create this "green miracle" on their emerald isle. They lowered taxes, instituted more favorable labor laws, supported university research, created incentives for companies to settle in Ireland, and developed a critical mass of international companies in related industries. For example, Microsoft sells all its software for Europe out of Ireland because of its favorable tax rates, which saves money for Microsoft and generates an additional $77 in new tax income for each Irish citizen.[34] Ireland adopted policies to become the "foremost knowledge-based society in Europe," and it is fast realizing this goal. And of course the country is endowed with important knowledge-age resources, such as excellent universities, hardworking people who speak English, an excellent location, a deserved reputation for hospitality, and a gloriously beautiful country to attract employees. When you visit Ireland for business, there is excitement and optimism. No wonder companies such as Dell, Microsoft, Oracle, Gateway, Hewlett-Packard, IBM, and Pfizer are flocking to Ireland. As a result, the Irish who left Ireland just a few years ago are going home. The Irish are entering the world of abundance.

People in only a few nations seem excited about innovation and competitiveness. We mentioned Ireland. Others include Singapore, Taiwan, India, and China. In spite of huge infrastructure limitatons, India is becoming the preeminent global supplier of information services. Wipro, Tata, Infosys, and Satyam Computer Services are multibillion-dollar knowledge-based service companies growing at 20 percent to 40 percent a year. If you visit Electronic City in Bangalore, you will find not only Indian companies, but a who's who of international companies, including HP, Motorola, 3M, Intel, GE, and Siemens.

China is becoming the preeminent global manufacturer, producing 50 percent of the world's cameras, 30 percent of the air conditioners and televisions, and much more.[35] The Chinese culture includes a prodigious work ethic plus legendary entrepreneurial zeal. Chinese managers are willing to work twelve hours a day to serve their customers. In addition, China's leadership is thoroughly committed to innovation. Remarkably, 100 percent of the eight members on the Chinese Politburo are engineers from Tsinghua University in Beijing.[36] The current Chinese president, Hu Jintao, was a hydraulic engineer. He has said, "Science and technology are the decisive forces in economic and social development in the world and innovation is a core part of a country's competitiveness." *Decisive* is a strong and accurate word. In a survey of Chinese and U.S. manufacturers by *Industry Week,* only 26 percent of U.S. respondents cited innovation as a top objective, as opposed to 54 percent in China.[37] China may not be a free country, but it has embraced many of the messages of this book. They have arrived.

Consider another incredible example: Eleven former Soviet bloc countries now have adopted flat-tax policies to reduce bureaucracy and waste—muda—and to encourage faster economic growth. Russia has a maximum rate of only 13 percent. We note wryly that this is an unusually practical form of socialism.

When you benchmark the United States against these global developments, you are left with the impression that the United States is still not fully serious about innovation and competitiveness. The relative success of the United States until now has allowed many of the underlying danger signs and weaknesses to be ignored, such as poor K–12 education. We are optimistic that the United States will respond, as it did during the 1980s to the challenge from Japan. But it will be a hard struggle, many will be unnecessarily hurt in the process, and a family of specific actions will be required to assure success.

Approach to Improving the Ability to Innovate

For developed countries innovation is the key to growth, prosperity, and quality of life. It is what will allow the United States to provide quality jobs and to be able to afford advanced health care and other social services. The understanding and broad application of innovation best practices across both the civilian and governmental communities can positively impact the United States's ability to thrive over the next century.

Because of their central importance for countries, there is a large body of knowledge devoted to the conditions that promote innovation, productivity, competitiveness, and job growth—which we consider to be innovation best practices. Many of these practices apply directly to countries and their governments, as well as to companies. Innovation and competitiveness within a country also constitute a discipline that can be learned and applied. Ireland proves that this is possible.

Like companies, however, very few countries embrace and use innovation best practices. If we reshape Gary Hamel's quote at the start of Chapter 16, we could say, "Every country's president will at least give lip service to the idea that the world is moving faster and that we need to do a better job at innovation and competitiveness. But if you go into a country and ask people to describe their innovation and competitiveness systems, you get blank looks. They have none." For those nations like Ireland who want to succeed by embracing innovation best practices, therein lies the opportunity.

Use of Innovation Best Practices by Government

What are examples of innovation best practices for countries? Some of the most important work over the past twenty years has been by Michael Porter of the Harvard Business School. Porter has pointed out the importance of "clusters" in creating an environment for economic growth. He defines clusters as "geographic concentrations of interconnected companies and institutions in a particular field. . . . [M]any clusters include governmental and other institutions—such as universities, standards-setting agencies, think tanks, vocational training providers, and trade associations."[38]

Porter uses the California wine industry as a comprehensible example. The wine industry "includes 680 commercial wineries as well as several thousand independent wine grape growers. An extensive complement of industries supporting both wine making and grape growing exists, including suppliers of grape stock, irrigation and harvesting equipment, barrels, and labels; specialized public relations and advertising firms; and numerous wine publications aimed at consumer and trade audiences. A host of local institutions is involved with wine, such as the world-renowned viticulture and enology program at the University of California at Davis, the Wine Institute, and special committees of the California state senate and assembly."

Silicon Valley is a cluster centered on various high-tech industries, such as computers, software, medical devices, and telecommunications. Hollywood is a cluster for movie production and entertainment; Wall Street in New York City is a cluster for financial services; and Washington, D.C., is a cluster for politics. Porter's work emphasizes the efficiencies and competitive advantages that come from proximity, flow of ideas, movement of employees, and a critical mass of necessary services. His work is also important because he demonstrates that the competitive advantages of clusters has been primarily local. This means that local governments can, through focused policies, help to increase their advantage, such as R&D support for universities, good transportation, and recruitment of complementary companies from around the world. Given the dramatic changes arising from globalization and high-speed communication, however, the definition of a cluster is being redefined, as many functions can now be offshored or outsourced. Increasingly, local clusters must emphasize the capacity to innovate.

Matty Mathieson directs a team at SRI that develops economic plans for regions and countries around the world. They have worked in more than 100 countries, helping to instill best practices that will lead to economic growth and prosperity. The basics are:

1. Strong *economic foundations*—human resources, technology and innovation, financial resources, hard and soft infrastructure, and a legal and regulatory environment. These foundations support and propel competitive industry clusters.

2. Dynamic *industry clusters*—a critical mass of firms in areas of a nation's comparative and competitive advantages. These can be high tech, medium tech, or low tech.

3. High or rising *quality of life*—which contributes to both economic foundations and competitive industry clusters.

Obviously, government policies—such as taxes, regulations, and labor laws—can either help or hinder these conditions and thus innovation and productivity. In each case there are innovation best practices that can increase growth.

Innovation Best Practices Within Government

Consider next governmental service, such as health care, tax collection, police, park services, or environmental protection. These agencies often work under severe restrictions and difficult political situations. In addition, civil service employment rules were obviously not designed to maximize efficiency and good management practices. Nevertheless, given these realities, when you interact with a government agency, how often do you feel that it's focused on customer value? Sometimes you probably do, and it may give you a positive feeling about the use of those tax dollars. The U.S. Postal Service, for example, has in some ways done a better job over the last decade, although it has a long way to go before it is competitive.[39] At other times, you probably want to scream because the government service is so poor—maybe it even seems hostile. What makes the experiences different? To a large extent it is a result of the organizational and management practices used.

Congress could help address these failings by mandating that all government agencies have a continuous improvement policy based on delivering the highest customer value using innovation best practices. Progress would have to be measured and quantified. Certainly, the Five Disciplines of Innovation can be applied to all government functions to aid in providing superior customer value. They can be initially applied to any group or larger function without requiring major reorganizations or modifications to work rules. They only require a change in how employees think about the work at hand—a change toward striving to provide continuously improved customer value. The goal is not to reinvent the government but, rather, to continuously improve it.

A simple way to start in a government agency is to ask, "Can each of our groups define who its customers are and prepare a quantitative NABC value proposition that conveys our contribution to them?" Other questions to ask include "Are we working on the most important issues for our customers or are we wasting our time on issues that are just interesting? Can we quantify our contributions to our customers and other stakeholders?" Our experience in working with many different government organizations is that they will not be able to answer these questions. The result of not being able to answer these questions is non-value-adding activities, inefficiency, and waste. There is a great deal of muda in most government organizations. Government employees often seem listless and dispirited because *they know* they are wasting their time and giving poor service. Letting these conditions continue is a disservice to the human spirit.

In addition to awareness of customer value, we have found that few government agencies have processes in place to continuously improve the value they provide to their customers. That is, they do not have a value-improvement process to continuously find, modify, and invent new innovative solutions for their customers. Most don't even think of that as a fundamental aspect of their jobs. Imagine Toyota not having a value-improvement process in place. You can't, because the company wouldn't stay in business. What is the difference between a gigantic organization like the IRS, which has roughly 100,000 employees, and Toyota with roughly 250,000? In some important ways, not much. They both have endless opportunities to innovate and create additional customer value. Colleen Kelley, president of the National Treasury Employees Union, recognizes the need for improvement to preserve jobs. She said, "The better course . . . is for agencies to work with their employees on continuous improvement of work processes."[40] But this can happen only if there are disciplined innovation processes in place.

Government does have positive examples of improvement. For example, the California Department of Motor Vehicles made a major improvement when it created a phone and Web service so that people could make appointments to get new driver's licenses, take driving tests, and register their cars. Earlier, you just showed up and waited your turn in line, which represented a huge amount of muda. The new system was one step toward better customer value, but since then the DMV seems to have stopped improving. What is required is a commitment to *continuous* improvement of customer value.

One excellent role model for the use of innovation best practices in government is Rick Baker, the mayor of Saint Petersburg, Florida. His objective is "to make Saint Petersburg the best city in America." Before he became mayor he developed a written plan for how the city could achieve this admirable goal. He had done his homework. Baker studied and visited fifty of the leading cities in the United States to learn best practices. His plan included the use of continuous improvement to constantly deliver greater value to his customers—the public. Among the initiatives he launched were performance metrics for all government services. The public can now track the actual performance achieved each year by looking on the city's website. Each metric is displayed in a standard, easy-to-understand graph. You can check and see, for example, that the time to fix sidewalks has gone from 14 months down to a few weeks. There are superb people like Mayor Baker to learn from to profoundly improve government services.

As part of a value-improvement process, agencies can identify all the organizational barriers that prevent them from delivering the highest customer value. It should be part of their jobs to work systematically to eliminate those barriers and achieve organizational alignment.

Finally, consider governmental research and innovation services. In the United States, the government is the primary source of funding for basic research. We discussed earlier how DARPA is a model for government-supported R&D. It is an example of innovation best practices. DARPA is customer-driven—to meet the needs of the Department of Defense it works on important problems, hires champions, assembles superb innovation teams, continuously iterates its value propositions, addresses the human issues, and creates organizational structures that are appropriate to the tasks at hand. This is, unfortunately, not a model that most of the funding organizations of the U.S. government use. We are sure that if they did, the effectiveness of the money spent would be significantly improved—perhaps by 20–100 percent per year, or more.

There is another way government research funding agencies could become more effective: They could directly support programs, organizations, and companies that exemplify innovation best practices. It is possible to imagine in the future the U.S. government saying, "We will give you this contract only if you have been certified as an organization that demonstrates the following benchmarks of Continuous Value Creation [CVC] success." Today this happens in software, where for certain projects you must be ISO 9001–certified to develop software for the government.[41]

Innovation in Education

Next, consider education, where the Five Disciplines can become the basis for a K–16 program. An appreciation that innovation is the fundamental driver of prosperity and quality of life should be integrated into the basic curriculum. The United States needs to retool education from kindergarten through graduate education, creating "an innovation culture at all levels, and providing students the opportunities to explore open-ended problems, engage in teamwork, and work on projects that cross traditional disciplinary lines."[42] The fundamentals of innovation and entrepreneurship, for example, can be taught as part of traditional K–12 classes, such as history, science, and social studies. For some students, a specific program based on innovation and entrepreneurship could be inspiring. Our colleagues at SRI have suggested that there should be an ISO standard for education.[43] ISO stands for the International Organization for Standardization, which develops standards for management processes as well as other areas of endeavor. The key feature of an ISO standard for education is that it would measure the *rate of improvement* of the school. Once an organization accepts continuous improvement based on delivering the highest customer value, positive change is almost assured.

Other countries have already reached these conclusions and taken action. Singapore, for example, teaches courses on "innovation and enterprise" to all students. Tharman Shanmugaratnam, minister of education for Singapore, announced, "[T]he key focus for the ministry of education . . . would be to nurture a spirit of innovation and enterprise [I&E] among our students and across our schools."[44] Singapore recognizes that the future of its small country of 4.4 million is dependent on a well-educated workforce equipped with new skills in creativity, teamwork, and human values, such as respect for others and integrity.[45] The country is creating a comprehensive program with new rewards for teachers, "360-degree" supervisor feedback, and continuous improvement. Is it any wonder that Singapore, on the "Innovation Index," is projected to move to the top of the index for emerging nations?[46]

Clearly, the United States needs to establish a "National Innovation Agenda" for our educational system.[47] We need to isolate the specific aspects of our educational system that need repair and move forward to enable a culture of innovation. This would embrace scientists and engineers at all levels and provide them with a collaborative, multidisciplinary curriculum that includes value creation and innovation.

Similar advances are needed in university education. We mentioned earlier the small, and declining, numbers of science and engineering students being graduated each year in the United States—only 7 percent of the number from India and China. Making science and engineering more interesting to U.S. students is a vexing cultural issue. Obviously, the endless stories about outsourcing do not help to make engineering more attractive. Demonstrating that the future is full of opportunities for those with the right skills would help. But there is another problem that can be more directly addressed: Students today can sense that in the exponential economy, a traditional science and engineering education is insufficient. They may not know what is missing but they are voting with their feet. In order to make science and engineering exciting again, we must add value creation to the curriculum. All engineering students should take a course in the fundamentals of innovation and entrepreneurship. And all students should be taking courses with an interdisciplinary focus, because "innovation tends to occur more frequently at the intersection of disciplines."[48] In the exponential economy, these are life skills that transcend most of the technical specifics of a traditional undergraduate education.

One university, Worcester Polytechnic Institute (WPI), has taken a huge step in these directions.[49] All students at WPI must complete two team projects to graduate. These three- to four-person teams work on projects at different locations around the world, including Ireland, Stockholm, Hungary, Hong Kong, Melbourne, Zurich, and India. This Global Perspectives Program helps them learn many of the fundamental lessons we describe in the Five Disciplines of Innovation. WPI is a role model that should be widely emulated.

Innovation in the Media

The Internet is a powerful mechanism to spread best practices about innovation and entrepreneurship. In addition, increasingly in the United States there are shows popping up that address issues surrounding entrepreneurship. For example, consider *Start-Up Nation,* a radio call-in show that discusses the issues of forming and running new companies. Such shows have a surprisingly large audience. Why surprising? Do you know that there are 5.7 million U.S. companies with fewer than 500 employees, which represent 99.7 percent of all businesses in the country? These are an important part of the economic engine for the United States economy. But the mainstream mass media currently provide little, if any, insight into the significance of these companies and their needs. Innovation and its importance to the future of the United States are given even less serious attention.

Over the next decade, the mainstream media will spend a great deal of time talking and writing about the consequences of our lack of competitiveness and its impact on big companies. They will be concerned when Chinese firms buy big United States companies, when software jobs move to India, and when protectionism becomes a political issue. But these stories, absent the bigger context of innovation and competitiveness, represent a disservice to their readers and listeners. The action is not just with the larger companies, which, as we discussed in Chapter 2, are finding it increasingly difficult to survive, but rather at the grassroots level, where new company formation through innovation thrives. The mainstream media are basically missing the story. Perhaps each media-outlet science and technology editor can change his or her job description to "Science, Innovation, and Entrepreneurship Editor" to more accurately reflect the reality of our situation.

More generally, the mainstream media can increase their active role in helping communicate the challenges, opportunities, and excitement of innovation and entrepreneurship in our time. They could help stimulate a productive discussion about how the United States needs to improve its schools, tax policies, government regulations, and research agencies. They too could use innovation best practices as a ruler to measure the United States' performance against that of others. Ultimately, they could help promote and exemplify the skills and attitudes needed to thrive in the exponential economy. For example, maybe the creative staffs of the major television networks could make this the ultimate "survivor" show.

One bright spot is a program at Stanford University called Innovation Journalism. It is not about innovation *in* journalism but, rather, it is a journalism program *about* innovation.[50] The program is led by David

Nordfors with the goal of advancing the public debate about this critically important topic. He has assembled journalists and students from many countries to be part of the program, who then become innovation-enlightened journalists at major publications around the world.

The entertainment industry can generally be described as ill-informed, or even antagonistic, to the process of innovation and its central role in creating a prosperous, safe, environmentally healthy nation. Under the heading of small comfort, there are now reality shows involved in related issues, such as *The Apprentice,* with Donald Trump, although these shows skip over the underlying issues of innovation and entrepreneurship. It might also be too much to ask of Hollywood, but it would be terrific if the good guys in a movie were occasionally entrepreneurs rather than having businesspeople always cast as the bad guys. In a rational universe, those good folks who have started successful companies would be treated like heroes—because they are.

We have given several examples of how innovation best practices can be used in different government and commercial organizations. "Best practice" often refers to the best current practice. Each enterprise needs to first develop a sound basis for evaluating new innovation best practices. This requires a clear understanding of what "value" means to all stakeholders: the customers, enterprise, investors, employees, and public. The sound basis also includes copying practices from any industry or organization deemed to exhibit current best practices. Exemplars of best practices might include J&J, Intel, Dell, Toyota, DARPA, W. L. Gore, and the Gates Foundation. Other countries are exemplars too, such as Ireland, Singapore, and China. But just as important, each organization and nation should develop practices that are even better and more appropriate to their setting than those of their counterparts in other countries.

Our experience has been that the use of innovation best practices is a positive stimulus for employees. The overwhelming majority of employees want to do a good job, learn new skills, and deliver greater value to their customers and other stakeholders. An enterprise-wide commitment to delivering the highest customer value based on innovation best practices is a unifying goal for all. This can be true of countries as well. We have given but a few examples—in government, education, and media—but every activity can benefit from the use of innovation best practices.

Benefits per Costs of Improving the Ability to Innovate

Focusing on the overarching themes of creating and delivering the highest customer value through the use of continuous improvement and innovation best practices has many benefits for a nation. A major objective in all nations is to create high-paying jobs. Even a small change in productivity from improved innovation processes can have a huge impact. For example, consider the United States with a GDP in 2005 of $12 trillion.[51] Assume that it continues to grow at its current rate of 4 percent per year. What would the impact be over the next ten years by going from 4 to 5 percent growth? That 1 percent change would represent an additional $1 trillion in GDP growth, an increase that would support more than 10 million new jobs paying an average of $30,000 to $50,000 a year each. In comparison, over the last five years the United States has created approximately 4 million new jobs.[52]

Let's turn this discussion around: Can we imagine the United States improving its innovation processes to realize an additional 1 percent yearly increase in GDP? First, we have seen companies embrace innovation best practices and go from negative growth to double-digit growth. A big impact could be made if even a fraction of U.S. companies embraced these ideas. Think of the "Big Three" automobile companies: General Motors (GM), Ford, and DaimlerChrysler. Over the past fifteen years they have lost roughly one percentage of market share annually.[53] If, instead, the Big Three reclaimed their market position, the impact on the U.S. economy would be significant, resulting in additional revenues of tens of billions of dollars a year. Multiply that by several thousand companies and there would be a positive impact worth many hundreds of billions of dollars per year.

Second, consider that 500,000 new companies are started in the United States each year. Small companies, with fewer than five hundred employees, employ roughly half the nation's workforce and produce half the commercial nonfarm GDP. They are responsible for 60 to 80 percent of the net new jobs each year and create fourteen times as many patents per employee as bigger companies. In 2001, they accounted for 97 percent of all exports.[54] In addition, while 75 percent of R&D expenditures comes mostly from firms with more than one thousand employees, these large firms are often not the genesis of paradigm-changing innovations. The R&D expenditures of large firms largely go toward improving their current product lines, while only 8 percent of their R&D is spent on major new innovations. It is within the smaller firms that the radical, paradigm-shifting innovations often occur. Small firms' patents are at least twice as likely to be found among the top 1 percent of high-impact patents.[55] And small firms are more "effective in producing high-value innovation."[56] It is certainly no overstatement to say that "the contribution of small businesses to the innovation economy rivals those of large ones." As Robert Litan, vice president of Research at the Ewing Marion Kauffman Foundation, says, "A disproportionate number of innovations are developed by entrepreneurs as opposed to big firms."[57]

Nevertheless, the failure rate of companies with fewer than twenty employees is striking. Different estimates suggest that only 30 to 50 percent of new companies survive four years after formation. Elizabeth Seanard and Lloyd Taylor[58] quote A. Garcia of the Small Business Administration, who reports that "small businesses increase their chance of survival by 50 percent by preparing a business plan." This is our experience too. Very few individuals either have the skills or do the homework necessary before forming a new venture. The waste here is huge, and a small improvement would pay great dividends.

Small companies have to overcome other hurdles that reduce their probability of success: Today, for example, more than sixty U.S. governmental organizations either regulate or impact small businesses. Some of these agencies help, but most represent costs to be absorbed. The Small Business Administration notes that the cost of compliance for companies with fewer than twenty employees is $7,000 per employee, which is 60 percent higher than for companies with more than 500 employees.[59] This is upside-down. We should be helping small companies, not dragging them down.

Supporting small businesses and increasing their survival rate and ability to innovate by a few percentage points seems like a modest goal. *Even that small improvement would eventually create millions of new jobs with salaries worth hundreds of billions of dollars per year.*

Third, does anyone doubt that the government could improve its own price performance by 1 percent a year? From our interactions with government services, improvements many times that estimate seem possible. The U.S. government's budget is more than $2.3 trillion dollars per year. Each 0.1 percent saved is worth $2.3 billion per year, which could be better spent on research, training, innovation, and job creation. But more important, large improvements in customer value can clearly be achieved without even cutting the government's budget.

Are these estimates even remotely possible and can they be achieved at a reasonable cost? Yes, we think so. Look at what Ireland has accomplished starting from a much lower economic level. Also, in most commercial organizations and government agencies, employees have little or no understanding of the most basic elements of customer value. In our workshop, for example, we ask senior management teams to write down the definition of customer value for their customers. We seldom find a management team that can give a coherent answer to this most basic of questions. This experience and many others convince us that enormous resources are being wasted. Remember Toyota. It went from making poor-quality cars after World War II to becoming the second-largest vehicle maker in the world. Toyota made these improvements not by increasing its costs but by reducing them. The previous chapter described how Toyota turned the worst automobile plant in America into the best in just a few years. Toyota is now the world's most profitable automobile company, with the resources to buy Ford, GM, and DaimlerChrysler. Toyota achieved all this by an uncompromising commitment to continuous improvement—which any organization or governmental agency can do too.

Finally, in an existential sense, the United States has no choice but to improve its innovation practices and competitiveness. National security in every nation is ultimately based on productivity, growth, and prosperity. Only by embracing innovation best practices throughout the nation will the United States maintain parity with the rest of the world and thrive.

Competition and Alternatives to Using Innovation Best Practices

There are a number of alternatives to using innovation best practices as a central, organizing theme to improve competitiveness. At the national level, some may argue that there are better approaches. For example, the old U.S. Industrial Age model, albeit extraordinarily exaggerated, was based on the idea that the boss knew best and therefore told everybody else what to do. This attitude was exemplified in the corporate world by CEOs like Henry Ford, Andrew Carnegie, and John Rockefeller.[60] In the exponential economy, these approaches can't work, because no matter how smart the leader is, a single person isn't remotely smart enough to create new customer value fast enough. Only by tapping into the genius of the entire organization can one possibly have an organizational IQ large enough for the company to thrive. This conclusion is true for nations as well as companies. Innovation best practices represent the family of ideas to achieve this.

Protectionism

Because the exponential economy results in continuous social and economic disruptions, many groups will fight hard to preserve the status quo. We are extremely sympathetic about the dislocations that are occurring economically around the world. Attempts at protectionist legislation are proposed almost daily. We also see attempts by unions to preserve jobs even at the expense of the survival of the companies where their members work. Union leaders need to see that if they are not aligned with management to create the highest customer value, there will be no jobs. We see other attempts to impose wage and price controls in response to increasing prices in commodities, such as automobile gasoline. Ultimately, these efforts are futile and counterproductive. Holding back the tsunami of change represented by the exponential economy is impossible without doing even greater damage to businesses and national economies. The most governments can do is smooth out the transitions and ease the pain of those who will be inevitably disrupted.

To us the debate is not really over the use of continuous improvement and innovation best practices. It is, after all, hard to argue for the opposite. The real debate is over which best practices to select and which stakeholders will gain from different approaches. We don't expect, for example, that the arguments about which system is best—free-market capitalism, socialism, or communism—will go away quickly. Each of these has its own value system. Clearly, we believe that free markets create the greatest value for all shareholders, including the public. Nevertheless, an open discussion about innovation best practices in each setting will lead to progress. The focus on innovation best practices, such as CVC, frames a more rational debate where one can begin to quantify value for all.

Consider a current example, China. China remains a communist dictatorship with fundamental limitations on individual freedom. Obviously, communist systems have serious restrictions that will eventually inhibit their ability to innovate. In the exponential economy, the free flow of information is the equivalent of a nation's life-giving irrigation system. But the Chinese leaders have learned important lessons from what happened in other communist countries, such as the Soviet Union. Many of our Chinese colleagues now joke that they are more capitalist than the West. And in some ways, they are. As we described earlier, Chinese leaders pay great attention to innovation best practices and attempt to shoehorn them into their system while preserving, to the extent they can, their communistic structure.

We have seen firsthand that the failure to use innovation best practices does enormous damage to both organizations and employees. The best solution for a country is to embrace the realities of the exponential economy and create the infrastructure, governmental policies, and culture required to thrive in it. A country that is rapidly growing will certainly exhibit dislocations. But a country that is rapidly growing will also create increasing, and better, opportunities for its people. The ability to work in a country that is full of opportunity, with prosperity and excellent quality of life, is the best alternative. That is why we advocate the extensive use of innovation best practices to ride the exponential economy wave as opposed to being drowned by it.

Conclusions

Many people see the world as one of scarcity, whereas we see the world as one of abundance. How many quality jobs are there, after all? Is there really an unlimited supply? We don't know how to answer this question, but we have a strong intuition that we are not remotely near the limits imposed by a fixed number of jobs.

We remember being in graduate school talking to our socialist friends about the limits of growth. They tried to convince us that the U.S. economy was going to grind to a halt because there were not enough jobs. At the time, we expressed skepticism that such a thought could be correct, since none of the things we wanted in our lives had yet been achieved. For example, there were no effective cancer therapies, infectious diseases were still killing millions of people around the world, and our ability to communicate was limited to ungainly, low-bandwidth analog devices. In addition, cars were polluting the environment, and government services were unresponsive and poor.

Now, decades later, and although our economy's GDP has grown from $2 trillion a year then to $12 trillion per year today,[61] we still don't have effective cures for most cancers, the Internet is an immature, unsafe communication system, we need nonpolluting automobiles, and government services haven't significantly improved.

Paul Romer of Stanford University has pointed out, "While things such as land, machinery and capital are scarce . . . ideas and knowledge are abundant and . . . they build on each other and can be reproduced cheaply or at no cost at all. In other words, ideas don't obey the law of diminishing returns, where adding more labor, machinery, or money eventually delivers less and less additional output."[62]

In addition needs keep expanding, as represented in Maslow's and Markowitz's hierarchies.[63] The Industrial Age, for example, was about making products at low cost. Deming and Ohno added quality to low cost. These two aspects of customer value are now expected. Increasingly, we are moving beyond low-cost, high-quality products. Customers are demanding new and better services. They are also demanding a positive experience that enhances their personal identity, as exemplified by Starbucks, Apple, and Lexus. This observation is not new—it is just becoming more ubiquitous. After all, Walt Disney understood the enormous value that comes from a great experience fifty years ago when he created Disneyland.[64] This transition from products to services to experiences to identity to deeper meaning creates the opportunity for millions of new job opportunities. For all these reasons, we're betting on a world of abundance full of remarkable opportunities for the foreseeable future, where improvements and progress in India and China can lead to improvements in the United States and vice versa. In many important ways, it will be win-win for the world.

Throughout this chapter, we have highlighted the many faults of the United States. America also has great strengths that can be leveraged to thrive in the exponential economy. It is remarkable, for example, that although the United States has spent many hundreds of billions of dollars on the war against terror, has assimilated tens of millions of poor immigrants, and has had to recover from the extravagances of the Internet bubble, the ravages of Hurricane Katrina in Louisiana, and the gas price shock, the U.S. economy is growing at a robust 4 percent, with the highest improvements in productivity in the past fifty years. This growth rate is two to three times greater than most of the economies of Europe and Japan. Americans are practical, optimistic, flexible, and future-oriented. They are adjusting rapidly to the realities of the exponential economy. America has the world's best capital markets and a culture of hard, focused work. Although we still have our share of racial, ethnic, and sexual discrimination issues to work through, we are in principle a nonhierarchical, egalitarian meritocracy. Most important, we are a free country.

Many of us remember the sixties and seventies, when the demise of the United States was predicted throughout the elite media. Those predictions were wrong. The advantages of nations with central planning were overestimated. The inherent strengths of America were underestimated, especially the ability to adapt. The ability to adapt will become an even bigger advantage as we move further into the exponential economy. We believe the predictions about the demise of the United States are wrong this time too. Clearly, America has critical issues that need to be addressed. But, by a process that will leverage its strengths, America will make the transition and remain a great nation with a strong economy.

We see this as a time for society to be empowered by an innovation imperative:

————————{ THE INNOVATION IMPERATIVE }————————

- We live in a world of abundance, not scarcity.
- Only through innovation can this abundance be converted to growth, prosperity, and quality of life.
- The role of teams, enterprises, and nations is to create the highest possible value for their customers and other stakeholders.
- The highest value is created using innovation best practices—the Five Disciplines of Innovation.

We hope you, your team and organization, and your country will join us in the exponential economy to solve the problems we face and to make a positive contribution to the world. There has never been a more important, exciting time. Be a champion, embrace the Five Disciplines of Innovation, and begin now.

appendix

value factor analysis:
quantifying customer value

"PRICE IS WHAT YOU PAY AND [CUSTOMER]
VALUE IS WHAT YOU GET."[1]
Warren Buffett

In Chapter 4 we presented a simplified version of Value Factor Analysis (VFA), comparing a Ford Taurus with a Toyota Prius. Figure 1 is an example of a complete spreadsheet, which you can use to facilitate Value Factor Analysis. While it looks a little overwhelming at first glance, the calculations are so simple they can be performed on a paper napkin over lunch. The spreadsheet just makes it easier to try out different assumptions and be more complete in your analysis.

SRI VALUE FACTOR ANALYSIS ™

QUALITY ATTRIBUTES	CUSTOMER Importance	COMPETITOR 1 Satisfaction	Benefit	COMPETITOR 2 Satisfaction	Benefit	OUR GLUE Satisfaction	Benefit
1. Electrical conductivity	5	5	25	4	20	5	25
2. Electrical shorting	5	4	20	4	20	5	25
3. Pitch limit	5	3	15	3	15	5	25
4. Thermal conductivity	5	4	20	3	15	3	15
5. Substrate adhesion	4	2	8	3	12	4	16
6. Thermal cycling	4	3	12	4	16	4	16
7. Moisture resistance	4	3	12	2	8	4	16
8. Aging	4	4	16	4	16	4	16
9. Appealing smell	3	3	9	4	12	4	12
10. Appearance	4	4	16	2	8	3	12
TOTAL QUALITY BENEFITS			**153**		**142**		**178**

CONVENIENCE ATTRIBUTES	CUSTOMER Importance	COMPETITOR 1 Satisfaction	Benefit	COMPETITOR 2 Satisfaction	Benefit	OUR GLUE Satisfaction	Benefit
1. Cure time	5	3	15	2	10	3	15
2. Material shelf life	2	4	8	2	4	3	6
3. Material storage temperature	2	4	8	2	4	5	10
4. Ease of application	3	3	9	4	12	4	12
5. Cleanup	2	3	6	3	6	4	8
6. Disassembly	2	4	8	2	4	2	4
7. Hazardous material disposal	5	1	5	4	20	4	20
8. Toxic Material—Safety	5	1	5	4	20	4	20
9. Service and support	4	4	16	3	12	4	16
10. Distribution	4	4	16	2	8	5	20
TOTAL CONVENIENCE BENEFITS			**96**		**100**		**131**

COST ATTRIBUTES	CUSTOMER Importance	COMPETITOR 1 Expense	Cost	COMPETITION 2 Expense	Cost	OUR GLUE Expense	Cost
1. Material cost	5	$	5	$$	10	$$$$$	25
2. Storage	4	$	4	$$$	12	$$$	12
3. Material unit cost	5	$$$	15	$	5	$$$$$	25
4. Cure and oven cost	3	$$	6	$$$$	12	$$$$	12
5. Labor	3	$$$	9	$	3	$$$$$	15
TOTAL COSTS			**39**		**42**		**89**
VALUE FACTOR = (QUALITY * CONVENIENCE)/COST			**377**		**338**		**262**

Figure 1: Value Factor Analysis spreadsheet for a new glue to be used in electrical applications.

The objective for a new product or service is to provide much greater customer value than the competition or alternatives. VFA is a tool to help you achieve this objective. VFA also aids in uncovering hidden customer needs and in better understanding the more subtle attributes of the competition.

In Chapter 4 we simplified the Value Factor by representing it as:

Value Factor = Benefits/Costs

But, as we indicated, it actually consists of four variables—quality and convenience and their respective costs. We consider these to be independent attributes, as indicated in Figure 4.1, which gives:

Value Factor = [(Quality benefits) × (Convenience benefits)] / [(Quality costs) × (Convenience costs)]

Separating out these different potential benefits and costs helps identify all the ways you can create new customer value. Convenience costs can also include all forms of inconvenience, the elimination of which can significantly increase customer value. If you have to travel one hundred miles to get your car repaired, for example, it significantly reduces the car's value for you. Finally, we note that an enterprise's ability to thrive in the exponential economy can be measured by how fast this quantity increases over time per dollar invested.

Using Figure 1 as an example will help you understand how to perform Value Factor Analysis for your product or service. We will simplify the analysis by assuming that the convenience costs = 1.

The example in the figure is for a new glue (Our Glue) to be used in commercial applications and where there are already two competing products. It is not important that you understand all the details about the glue, but that you see how VFA is calculated. It is an example that also demonstrates how VFA can help uncover the large number of attributes important to the customer. As a rough guide, a new product or service should have a Value Factor that is two to ten times greater than the competition for the difference to be noticeably significant.

Next we will go through the specific steps needed to calculate the VFA:

1. *List the Product's or Service's Attributes:* In Figure 1, note that in the leftmost column there are three categories—*Quality, Convenience,* and *Cost.* Start at the top of the figure in the first column and scan down it. Quality has ten attributes in this example, beginning with "electrical conductivity" and ending with "appearance." Convenience also happens to have ten attributes, beginning with "cure time" and ending in "distribution." Cost has five attributes, ranging from "material cost" to "labor." Of course the number of Quality, Convenience, and Cost attributes will vary depending on the product you are evaluating. Make sure you write down everything you can think of. If you write down less than ten Quality and Convenience attributes you are not trying hard enough.

2. *Determine the Importance of Each Attribute to the Customer:* Now go back to the top of the figure. You will see a column for *Customer Importance.* This column indicates how significant or "important" each product or service attribute is to the customer. In our example, "electrical conductivity" is rated 5 in importance using a six point scale, where 0 = no importance and 5 = most important. Note that since this is a commercial application, the customer attribute of "appealing smell" has an importance of only 3. By contrast, if we were evaluating a glue for home use, it would probably have an importance of 5. Next determine the importance ratings for all the attributes under Quality, Convenience, and Cost. To get started, these ratings can be estimated by you and your team. However, you will need to show your VFA to customers to obtain more realistic values. Pictures, mockups, prototypes, or samples are very useful during these discussions. Often other critical attributes will often be discovered, which had been previously overlooked. Also, don't forget subjective factors, such as smell, feel, and appearance.

3. *Performance of Each Attribute:* We next evaluate how each product's performance satisfies the identified Quality, Convenience, and Cost attributes. Under the *Satisfaction* columns we list how our customers would evaluate each product's attributes, again on a six point scale from 0 to 5, where 0 = no satisfaction and 5 = the highest satisfac-

tion. For example, look again at "electrical conductivity," which for Competitor 1 has a customer satisfaction score of 5. For Competitor 2 the satisfaction score is 4 and for Our Glue the score is 5.

4. *Total Scores for Quality, Convenience, and Cost:* Next under the *Benefits* column we compute the customer Benefits by multiplying each attribute's Importance times its Satisfaction. For example, using the quality attribute of "electrical conductivity," Importance has a score of 5 and Satisfaction has a score of 5, so the Benefits are $5 \times 5 = 25$. A similar calculation is done for all the Quality and Convenience attributes and for each product. Then the Quality Benefits and Convenience Benefits are all added to create the *Total Quality Benefits* and *Total Convenience Benefits* for each. For example, the Total Quality Benefits for Competitor 1 are 153.

   For *Cost* attributes the totals are computed similarly. The ratings for Expense are like ratings in travel books for restaurants, where $0 =$ no cost and $\$\$\$\$\$ =$ highest cost. For example, the first component, "material cost," is rated 5 for importance. Thus the Cost for Competitor 1, with one $ is $5 \times 1 = 5$. Again we add up the individual Costs to get the *Total Costs* for each product. For Competitor 1 the Total Costs are 39.

5. *Value Factor Calculation:* You are now ready to compute the VFA for each product, which is given by Value Factor = (Total Quality Benefits × Total Convenience Benefits) / Total Costs. Simply take the Total Quality Benefit score (Our Glue = 178) times the Total Convenience Benefits (Our Glue = 131) and then divide by Total Costs (Our Glue = 89) to obtain the Value Factor (Our Glue = 262). Note that we multiply Quality Benefits times Convenience Benefits because they are independent variables.[2]

The score for Competitor 1 is 377, and for Competitor 2 it is 338. Our Glue has a score of only 262. We conclude that our new product is probably not worth pursuing without significant improvement—its customer value is not higher than the competitors' customer value. But if we could produce our glue at a much lower cost, this conclusion could change. The alternative strategy would be to dramatically improve the quality and convenience benefits.

A recent retail product, Gorilla Glue, is such an example. It costs more than Elmer's and the epoxies that are currently available. But the satisfaction provided by its attributes is much higher than that of the competition. When you go to the hardware store, your might hear someone say, "Well, Gorilla Glue is more expensive, but I use it and find that it never fails. It is just phenomenal because it works on everything."

Value Factor Analysis is a simple tool to help facilitate analysis of the relative customer benefits per costs of one product compared with those of another. The results are approximate, but they give surprisingly good guidance. And Value Factor Analysis keeps everyone focused on improving value to the customer when compared with the competition.

Use Value Factor Analysis to anticipate improvements from your competition. As you are bringing your product to market, your competitors will still be improving their products too. The competition does not stand still.

There are other important design methodologies that can help you understand and improve customer value, such as Quality Function Deployment (QFD).[3] QFD includes a family of tools for planning and developing high-quality products based on including the "voice of the customer." We recommend that you read about QFD and add these ideas to your customer value creation toolbox. At SRI, even though we use a variety of tools, Value Factor Analysis is used on all projects, because it is a simple tool that makes *customer value* the primary focus.

gLossary

BENEFITS: Those features or attributes of a product or service that have value to a customer. Benefits have a financial equivalent—and they come in many forms, as services, convenience, experiences, emotions, and identity.

BUSINESS MODEL: The way an enterprise will make money from an innovation. For example, the oldest model is the "shopkeeper," where a product or service is sold directly to the consumer. Other models include the "razor-and-blades" model, where a product, such as a razor, is sold at a loss and profit is made by selling numerous copies of a disposable part, such as shaving blades.

CHAMPION: An individual who has the passion, value-creation, and human skills needed to make an important innovation happen. Champions use and exemplify the Five Disciplines of Innovation.

CONTINUOUS VALUE CREATION (CVC): The Five Disciplines of Innovation when applied to and used by all functions of an enterprise. Other best practices complement and fit within CVC.

CUSTOMER VALUE: Customer value = Customer benefits − Customer costs.

DNA OF CHANGE: The three prerequisites for people to change. People must have a *desire* to change, see a *new vision* for the change, and understand the *action steps* required to make the change.

ELEVATOR PITCH: A short, memorable, and convincing presentation with a hook to gain interest, a core (i.e., a short NABC value proposition), and a close to ask for next steps.

EXPONENTIAL ECONOMY: Those knowledge-based products and services that improve in price performance at rapid, exponential rates. Examples include computers, communications, consumer electronics, and biotechnology. As more of the economy becomes knowledge-based, those sectors will enter the exponential economy too.

FIVE DISCIPLINES: Success in the exponential economy requires mastering innovation best practices in five different disciplines, all directed at the objective of rapidly creating new customer value. The Five Disciplines are (1) important customer and market needs, (2) value creation, (3) innovation champions, (4) innovation teams, and (5) organizational alignment.

FUD: The most common psychological blockages to change, which are *fear, uncertainty,* and *doubt*.

INNOVATION: The creation and delivery of new customer value in the marketplace that also provides a sustainable return to the enterprise.

INNOVATION PLAN: A written document based on a value proposition plus all the other ingredients necessary for success, such as market positioning, business model, risk analysis, intellectual property, financial plan, product schedule, staffing, and ROI.

INNOVATION TEAMS: Champion(s) and team members using and exemplifying the Five Disciplines of Innovation.

ITERATE: A shorthand way of saying, "Get new input and ideas, synthesize these ideas to improve your value proposition, and then repeat." Many iterations are needed to make a value proposition compelling.

LAW OF EXPONENTIAL INTERCONNECTIONS: The number of unique interconnections between users in a fully connected network grows exponentially as the number of users goes up. The same effect applies to teams: it is a metaphor for the potential power and difficulties that result from teams.

MARKET POSITIONING: A plan for introducing a product or service into the marketplace. The plan identifies the specific customers in a market segment, how the product will be sold to them, and how the product will be described and advertised to them.

MOORE'S LAW: Gordon Moore observed that computer power grows exponentially, doubling in performance at the same price roughly every 18 months. This is an example of the Principle of Knowledge Compounding.

MOTIVATION MANTRA: A workplace where people experience, achievement, empowerment, and involvement.

MUDA: A Japanese word for *waste.* It means non-value-adding activity.

NABC: See *value proposition.*

PERSONAL TRANSFORMATION: The process by which an individual can successfully move from one vision to a new or larger vision by leveraging their strengths in the new vision.

PRINCIPLE OF KNOWLEDGE COMPOUNDING: Knowledge-based activities that exhibit exponential improvement over time, where four conditions are satisfied: (1) an important customer and market need exists, (2) new ideas are available, (3) a way to compound these ideas exists, where one idea builds upon another, and (4) the necessary human, financial, and other resources are available to drive the process. The most famous example is Moore's Law, but there are many others.

SELLER VALUE: Seller value = Seller price − Seller cost.

THREE-LEGGED STOOL OF COLLABORATION: The three conditions necessary for teams to form and team members to collaborate. They are *shared vision; unique, complementary skills;* and *shared rewards,* all held together by collaborative, respectful communication.

TRUST COMMANDMENTS: The cornerstones of trust and cooperation in any business endeavor. They are *respect, integrity,* and *generosity of spirit.*

VALUE CREATION: A process that starts when a champion addresses an important customer and market need, writes down a value proposition, presents it in a Watering Hole, and then carries it forward to a full innovation plan and a successful project or venture in the marketplace.

VALUE FACTOR ANALYSIS (VFA): A comparison of the customer Benefits/ Costs for a proposed new product or service when compared with those of the competition and the alternatives.

VALUE PROPOSITION (I.E., AN NABC): The fundamental ingredients of every Elevator Pitch and innovation plan. They are (1) the important customer and market *need,* (2) the product or service *approach,* (3) the *benefits per costs* resulting from the approach, and why these are superior to (4) the *competition* and alternatives. These ingredients define the value propositions required for every project: at least one is required for the customer and another for either the enterprise and/or investors. Additional value propositions are required for the other stakeholders—partners, shareholders, employees, and the public. Effective value propositions are quantitative.

WATERING HOLE: An enterprise-wide meeting held every two to eight weeks where a community of people with a variety of perspectives and common interest in a market segment come together. Its purpose is to help people rapidly improve their value propositions, proposals, and innovation plans; get access to new ideas and enterprise-wide resources; and learn.

WORTH: The sum of the customer benefits for a product or service, which has a monetary equivalent. An audio record, for example, might have a worth of $20 and a price of $10. The customer value would be the worth minus the price, or $10.

notes

In several places in the book, the names and other identifying details of individuals have been changed to protect the privacy of those individuals.

WHY LISTEN TO US?

1 Donald L. Nielson, *A Heritage of Innovation: SRI's First Half Century* (Menlo Park, CA: SRI International, 2005).

2 Of the 10 Emmys SRI's subsidiary the Sarnoff Corporation has won, 3 of them were when Sarnoff was part of SRI and 7 were when Sarnoff was part of RCA.

3 In 2005, Nuance was merged into the other leading speech company, ScanSoft. The combined company retained the name Nuance.

4 Failure rates depend on the particular industry, product, and service being referenced. An ACNielsen study conducted in Europe in 1997 concluded that the failure rate for new products (there were 525,000 launched in 1997) is 90 percent within two years. See http://acnielsen.com. Albert Page and Abbie Griffin, "The PDMA Success Measurement Project: Recommended Measures for Product Development Success and Failure," *The Journal of Product Innovation Management,* Vol. 13, No. 6, November 1996, pp. 478–96. "In the consumer packaged goods industry, the failure rate of new products can be as high as 90 percent. In the rest of the market, the success rate for new products is about 55 to 60 percent."

CHAPTER 1: THE ESSENCE OF INNOVATION: HOW FRANK HIT A HOME RUN

1 Paul Romer, personal communication, November 29, 2005. Obviously the application of capital and labor also drive growth and prosperity, especially in underdeveloped countries. We take that as a given; it is not the focus of *Innovation.*

2 The Centre for Innovation Studies at http://www.thecis.ca gives other useful definitions for *innovation.* For example, "We use the term *innovation* to mean the process

that transforms ideas into commercial value. A more systematic definition includes the:

1. Introduction and commercial sale of a new or improved product, e.g., the Pentium chip, Remington typewriter, tungsten filament light bulb, nuclear weapons, hybrid seeds, genetically modified foods.

2. Introduction and commercial use of a new method of production, e.g., Henry Ford's production line, the Pilkington float glass process, the Unipol polyethylene process, the steam engine.

3. Introduction of a new form of business organization, e.g., franchising, co-operatives, joint ventures, and co-production and outsourcing agreements, discount grocery stores, e-business, just in time manufacturing.

4. New uses for existing products, e.g., the electronic computer found its first uses in military applications such as calculating artillery trajectories. Its uses later broadened to include many more applications.

5. New markets for existing products, e.g., the donut was invented in Germany and subsequently spread throughout the world.

6. New distribution channels, e.g., the Internet is a new channel for selling books."

3 Daniel Stashower, *The Boy Genius and the Mogul: The Untold Story of Television* (New York: Broadway Books, 2002).

4 In addition to Frank Guarnieri and John Kulp, many other colleagues made critical contributions in the formation of Locus, especially Carmen Catanese and Vince Endres. The initial funding for formation came from Duane Mason of Prism Ventures, who invested $6 million because he trusted Carmen and Vince.

5 Locus Pharmaceuticals can be located at http://www.locusdiscovery.com.

6 From the Toyota employee handbook, "Benchmark Toyota Production System," no date.

7 Type "creativity" into the Web and you will find many consultants teaching different approaches to creativity.

8 "Microsoft's Midlife Crisis," *Business Week Online,* April 19, 2004, at http://www.businessweek.com.

9 Malcolm W. Browne, "Supply Exceeds Demand for Ph.D.s in Many Science Fields," *New York Times,* July 4, 1995.

10 Find additional information about Deming at http://www.deming.org. The quality movement in Japan was led by many individuals, such as J. M. Juran, K. Ishikawa, and G. Taguchi. See also Phil Cohen, *HCi* at http://www.hci.com for an overview of Deming's 14 points. Cohen writes, "Although Deming does not use the term Total Quality Management in his book, it is credited with launching the movement. Most of the central ideas of TQM are contained in the book *Out of the Crisis. . . .*

Variation was seen by Deming as the disease that threatened US manufacturing. The more variation—in the length of parts supposed to be uniform, in delivery times, in prices, in work practices—the more waste, he reasoned. From this premise, he set out his 14 points for management, which we have paraphrased here:

1. Create constancy of purpose towards improvement. Replace short-term reaction with long-term planning.

2. Adopt the new philosophy. The implication is that management should actually adopt his philosophy, rather than merely expect the workforce to do so.

3. Cease dependence on inspection. If variation is reduced, there is no need to inspect manufactured items for defects, because there won't be any.

4. Move towards a single supplier for any one item. Multiple suppliers mean variation between feed stocks.

5. Improve constantly and forever. Constantly strive to reduce variation.

6. Institute training on the job. If people are inadequately trained, they will not all work the same way, and this will introduce variation.

7. Institute leadership. Deming makes a distinction between leadership and mere supervision. The latter is quota and target based.

8. Drive out fear. Deming sees management by fear as counterproductive in the long term, because it prevents workers from acting in the organization's best interests.

9. Break down barriers between departments. Another idea central to TQM is the concept of the 'internal customer,' that each department serves not the management, but the other departments that use its outputs.

10. Eliminate slogans. Another central TQM idea is that it's not people who make most mistakes—it's the process they are working within. Harassing the workforce without improving the processes they use is counterproductive.

11. Eliminate management by objectives. Deming saw production targets as encouraging the delivery of poor-quality goods.

12. Remove barriers to pride of workmanship. Many of the other problems outlined reduce worker satisfaction.

13. Institute education and self-improvement.

14. The transformation is everyone's job."

[11] The research on lawsuits and physician-patient relationships is compelling. See, for instance, "Talking May Be Best Preventative Medicine," *New Jersey Law Journal,* February 28, 2005; Jordan Dolin and Theresa N. Essick, "Failure to Communicate: Insurers Can Help Reduce the Burden of Medical Malpractice Lawsuits by Urging Their Policyholders to Improve Physician/Patient Relationships," *Best's Review,* December 2004: p. 92(4); and Charles S. Lauer, "To Err Is Human; and a Heartfelt

Apology Might Be Just What the Doctor Ordered to Head Off Lawsuits," Publisher's Letter, *Modern Healthcare* (May 24, 2004): p. 20.

CHAPTER 2: INNOVATE OR DIE: THE EXPONENTIAL ECONOMY

[1] Steve Barth, "Champion of the Future Factory: A Conversation with Curt Carlson, CEO of SRI International," *Knowledge Management* at www.destinationkm.com, December 20, 2000. See also Stephen Abram, "Filling That Ever-Expanding Reservoir of Knowledge—A Report on the KM World 2000 Conference," *KM World,* December, 1, 2000. And, see Curtis R. Carlson, "What's the Value Proposition? Exponential Teams, NICs, etc." Engelbart Colloquium at Stanford, March 2, 2000, at http://www.bootstrap.com.

[2] See Reference.com at http://www.reference.com. Various versions of the quotation are found at different sites.

[3] Michael Kanellos, "Million-Resume Madness in India," CNET News.com, May 4, 2005.

[4] See CBS News, "Imported from India," June 22, 2003, on the Web at http://www.cbsnews.com. "With a population of over a billion people in India, competition to get into the IITs is ferocious. Last year, 178,000 high school seniors took the entrance exam called the JEE. Just over 3,500 were accepted, or less than two percent. Compare that with Harvard, which accepts about 10 percent of its applicants." "The IITs probably are the hardest school in the world to get into, to the best of my knowledge," says Vinod Khosla, who got into IIT about 30 years ago. "It's a big deal in India, it is," says Narayana Murthy, founder of the huge software company Infosys. He's known as "the Bill Gates of India." "Murthy's own son, who wanted to do computer science at IIT, couldn't get in. He went to Cornell, instead. Imagine a kid from India using an Ivy League university as a safety school. That's how smart these guys are." "I do know cases where students who couldn't get into computer science at IITs, they have gotten scholarships at MIT, at Princeton, at Caltech," says Murthy. In addition, entrance into IIT is based only on a competitive exam. Students enter Harvard on the basis of many criteria, including whether their parents went there.

[5] The decline in computer science Ph.D.s comes from the 2002–2003 *Taulbee Survey,* which can be found at http://www.cra.org/statistics/. The job-growth number comes from the National Science Board report of 2003, which can be found at http://www.nsf.gov/nsb/documents/reports.htm.

[6] "It is estimated that there are over one million lawyers in America—give or take a few thousand. The United States represents only 5 percent of the world's population, yet it is estimated that we have between 70 to 75 percent of the world's total number of lawyers—about one lawyer for every 300 individuals. Supreme Court Justice, Charles Evans Hughes said, 'The United States is the greatest law factory

the world has ever known.'" On the Web at http://www.lawyerethics.org/mt/archives/001215.html.

7 Alan Katz, "French Police Arrest 186; Paris Riots Continue for 10th Night," Bloomberg.com, at http://www.bloomberg.com.

8 Kathleen Madigan and Michael Mandel, "Commentary: Outsourcing Jobs: Is It Good or Bad?," *Business Week Online,* August 25, 2003.

9 Thomas L. Friedman, *The World Is Flat: A Brief History of the Twenty-first Century* (New York: Farrar, Straus & Giroux, 2005).

10 *Ibid.*

11 The Computer History Museum has a description of the origin and dissemination of Moore's Law. See http://www.computerhistory.org/about/tour/. In addition, for background information about the era, see John Markoff, in *What the Dormouse Said: How the 60s Counterculture Shaped the Personal Computer* (New York: Viking, 2005). See also John Markoff, "It's Moore's Law, but Another Had the Idea First," *New York Times,* April 18, 2005. "Mr. Moore was not the only one—or even the first—to observe the so-called scaling effect that has led to the exponential acceleration of computing power that is now expected to continue at least for the next decade. Before Mr. Moore's magazine article precisely plotted the increase in the number of transistors on a chip, beginning with 1, the computer scientist Douglas C. Engelbart had made a similar observation at the very dawn of the integrated-circuit era. Mr. Moore had heard Mr. Engelbart lecture on the subject, possibly in 1960."

12 Various estimates can be found for the computing power of the human brain, from 10^{14} to 10^{16} bits per second. By 2020 to 2025, this level of performance may be available for roughly $1000. See Hans Moravec, "When Will Computer Hardware Match the Human Brain?," *Journal of Evolution and Technology* 1, 1998, at http://jetpress.org/volume1/moravec.htm.

13 For supercomputer performance see "TOP 500 Supercomputer Sites" at http://www.top500.org/.

14 Moravec, *op. cit.,* pp 94.

15 Ray Kurzweil, *The Singularity Is Near: When Humans Transcend Biology* (New York: Viking, 2005), p. 136. In 2045, he predicts a "singularity" where "[t]he non-biological intelligence created in that year will be a billion times more powerful than all human intelligence today."

16 Emulating "human" performance is often a contentious topic. The architecture of the human brain is optimized to perform activities that are important to our survival. Human vision, for example, works efficiently because of a family of strategies that dramatically reduce the amount of processing required. The most obvious is the eye's fovea, which reduces the input data that must be processed by the

brain by approximately a thousand times: See C. R. Carlson and R. Klopfenstein, "Theory of Shape-Invariant Imaging Systems," *Journal of the Optical Society of America* A 1, 1040–1053, 1984. The difficulties of emulating and then surpassing human capabilities will eventually be mastered and included in future computing systems, but history shows that human capabilities are much harder to emulate than many researchers had initially thought.

[17] Kurzweil, *op, cit.,* p. 81.

[18] One evolutionary theory called "punctuated equilibrium" suggests there are times when little or no changes occur in a species evolution and then suddenly a mutation rapidly changes the species. Punctuated equilibrium helps explain why the fossil record does not gradually change. See S. J. Gould and N. Eldredge, "Punctuated Equilibria: The Tempo and Mode of Evolution Reconsidered," *Paleobiology 3,* 1977, pp. 115–151.

[19] It is not just exponential change that is hard for people to understand. Even linear growth is hard. See Arnold Kling, "What Causes Prosperity," *TCS Daily,* February 18, 2006, at http://www.tcsdaily.com/article.aspx?id=120302A. Most of the time it seems easy to believe that we were evolved for little or no change—homeostasis. See also Kurzweil, *op. cit.,* p. 10.

[20] The amount of rice needed was $(2^{64}-1)$ grains.

[21] "Colliding exponentials" is a term our colleague Pat Lincoln uses to get people to pay attention to those situations where exponential developments intersect one another. When this happens, change is rapid and often disruptive. An example is gene sequencing. The ability to sequence genes is increasing at 100 percent every 23 months because of the advances in both computers and gene-sequencing chips. See Kurzweil, *op. cit.,* p. 73.

[22] Bill Lehr and Frank Lichtenberg, "Information Technology and Its Impact on Productivity: Firm-Level Evidence from Government and Private Data Sources," 1977–1993, *Canadian Journal of Economics,* Vol. 32, No. 2, April 1999, pp. 335–62.

[23] These figures and estimates change rapidly. See "Internet Usage Statistics," Internet World Statistics, at http://www.internetworldstats.com/stats.htm. At the end of 2005, 15.7% of the world's population used the Internet.

[24] See Wikipedia, "Global System for Mobile Communications," at http://en.wikipedia.org/wiki/Global_System_for_Mobile_Communications.

[25] Federal Communications Commission, "Digital Television: FCC Consumer Facts," at http://www.fcc.gov. "Television stations serving all markets in the United States are airing digital television programming, although they can continue providing analog programming until February 17, 2009. At that point, broadcasting on the current (analog) channels will end and most of that spectrum will be put to other uses."

[26] Kurzweil, *op. cit.,* p. 74.

[27] Steven J. Spear, "The Health Factory," *New York Times*, August 29, 2005.

[28] Aubrey de Grey, "We Will Be Able to Live to 1,000," BBC News at http://news. bbc.co.uk/1/hi/uk/4003063.stm, December 3, 2004. Geneticist Aubrey de Grey from Cambridge University in England believes life expectancy will soon extend to a thousand years. You can imagine how controversial this claim is, but serious research is under way to treat aging as a curable disease. See also Paul Rincon, "Retirement Age 'Should Reach 85'," BBC News, February 17, 2006, at http://news.bbc. co.uk/1/hi/sci/tech/4726300.stm.

[29] "Coal Fuel Cell Has Promise," *Red Herring*, November 14, 2005. "Fuel cells are being looked to as a clean, inexpensive source of energy. Most fuel cells are seen as a potential replacement for batteries but coal-based fuel cells could have broader applications, producing general electricity for utilities. By adding oxygen to carbon in an electrochemical process, the direct carbon fuel cells (DCFCs) convert coal into electricity without burning it or turning it into a gas. The method can also use tar, biomass, and organic waste. The result is that twice as much energy can be produced from the same amount of fuel, at 20 to 30 percent lower cost and about half the carbon dioxide emissions, said Larry Dubois."

[30] "Quanta Computer Inc. to Manufacture $100 Laptop," Digital Opportunity at http://www.digitalopportunity.org.

[31] Arnold Kling, "The Most Important Economic News of the Year," TCS Daily, December 29, 2005, at http://www.tcsdaily.com/article.aspx?id=122805C.

[32] Kurzweil, *op. cit.*, p. 107.

[33] Nick Schultz, "The Great Escape: A Conversation with Nobel Prize Winner Robert Fogel," in TCS Daily, December 1, 2005, at http://www.tcsdaily.com. Also see William W. Lewis, *The Power of Productivity* (Chicago: University of Chicago Press, 2004) for a lucid discussion of the importance of productivity in improving the world's economic development.

[34] Richard Foster and Sarah Kaplan, *Creative Destruction: Why Companies That Are Built to Last Under Perform the Market—And How to Successfully Transform Them* (New York: Currency Publishers, 2001).

[35] Peter Schwartz, *The Art of the Long View: Planning for the Future in an Uncertain World* (New York: Currency Publishers, 1996); Peter Schwartz, Peter Leyden, and John Hyatt, *The Long Boom: A Vision for the Coming Age of Prosperity* (New York: Perseus, 1999).

[36] This phrase was coined by Joseph Schumpeter in his classic 1942 book detailing how innovative enterprises replace established companies. See Joseph Schumpeter, *Capitalism, Socialism, and Democracy* (New York: Harper Perennial, 1962).

[37] Charles Darwin, *The Origin of Species* (New York: Gramercy, Random House, 1979). It was originally published in 1859.

[38] In the book *Good to Great* (New York: HarperCollins, 2001), Jim Collins discovered

a small family of companies that outperformed the market over periods of fifteen or more years. These companies had common characteristics, which included: 1) leaders with personal humility and professional will, 2) hiring the right people as the first step ("get the right people on the bus, get the wrong people off the bus, and get the right people in the right seats"), 3) confronting the brutal business realities without losing faith, 4) staying with what you are passionate about, what will drive your economic engine, and what you can be the best at, 5) a culture of discipline, 6) thoughtful use of technology, and 7) relentless improvement (like "pushing a flywheel"). Collins's research and important insights show that companies can, under extraordinary conditions, outperform the market for decades.

[39] Jack Welch and Suzy Welch, *Winning* (New York: Collins, 2005); Jack Welch and John A. Byrne, *Jack: Straight from the Gut* (New York: Warner Business Books, 2001).

[40] Andrew S. Grove, *Only the Paranoid Survive: How to Exploit the Crisis Points That Challenge Every Company* (New York: Currency, 1999).

[41] See the Museum of Broadcasting Communications website for a history of broadcasting at http://www.museum.tv/archives/etv/S/htmlS/seeitnow/seeitnow.htm.

[42] Sarnoff Corporation was the silent partner who developed the digital video technology that made this innovation possible. The system employed a pre-standard version of MPEG-2.

[43] See "Digital Television" at http://history.acusd.edu/gen/recording/television2.html.

[44] For a discussion of different business models, see Wikipedia, at http://en.wikipedia.org/wiki/Business_models. It lists and discusses many models, including the subscription business model, razor-and-blades business model ("bait and hook"), multilevel marketing business model, network effects business model, monopolistic business model, cutting-out-the-middleman model, auction business model, bricks-and-clicks business model, loyalty business model, collective business model, industralization of services business model, low-cost carrier business model, and online content business model.

[45] The formula for compound interest is Next Year's Money = (Today's Money) × (1 + percent interest rate). Thus, the formula for the law of exponential improvement is Next Product = (Today's Product) × (1 + percent improvement). The difference from compound interest is that "percent interest rate" is replaced by the "percent improvement" due to new ideas that are added with each iteration of the process.

[46] There is no guarantee that the compounding rate for a specific activity will be constant over time, since it depends on the availability of new ideas and on many other factors, as noted. But our qualitative arguments require this to be only roughly true. We also note that the speeding up of the exponential economy seems assured as several billion people begin to participate more completely around the world,

which will provide more innovative ideas and drive increased competition. In addition, as the Internet and ways to use it become more ubiquitous, new ideas will be available faster and be compounded more rapidly to further increase the rate of price performance through new innovations. For example, there will soon be intelligent software "agents" working twenty-four hours a day over the Internet on our behalf, finding and sorting information for our use. Eventually, these agents will be hundreds of times "smarter" than humans and they will interact with one another at the speed of light to create, share, and compound new innovative ideas.

[47] Kurzweil, *op. cit.,* pp. 69–70. Kurzweil provides an explanation for Moore's Law and why it should speed up, which he calls the Law of Accelerating Returns. The interested reader should consult this fascinating and provocative book.

[48] Lewis Carroll, *Through the Looking-Glass* (New York: Dover Publications, 1999).

[49] Here is what the World Bank says about poverty in China: "Between 1981 and 2001, the proportion of population living in poverty in China fell from 53 percent to just 8 percent." Consult their website at http://econ.worldbank.org. See also Pamela Bone, "And Now for the Good News," *Business Week Online,* January 15, 2005, at http://www.businessweek.com.

CHAPTER 3: WORK ON IMPORTANT CUSTOMER AND MARKET NEEDS: THE RFID TAG

[1] The team members included Sunity Sharma, Larry Dubois, Philip Von Guggenberg, and Peter Marcotullio.

[2] Independent market analysts predict that the market for RFID applications is likely to exceed US $6 billion worldwide by 2010. See http://www.researchand markets.com/reports/c23616.

[3] *Ibid.*

[4] The BioX center at Stanford is a multidisciplinary program that works on a variety of initiatives, such as bio-computation, bio-sensing, bio-synthesis, and bio-storage; see http://biox.stanford.edu.

[5] Lisa Rosetta, "Frustrated: Fire Crews to Hand Out Flyers for FEMA," *Salt Lake Tribune,* December 14, 2005. See also http://www.sltrib.com/search/ci_3004197.

[6] "Defense Advanced Research Projects Agency," Wikipedia, at http://en.wikipedia. org/wiki/Defense_Advanced_Research_Projects_Agency.

[7] See DARPA at http://www.darpa.mil.

[8] Mathew Josephson, *Edison: A Biography* (New York: Wiley, 1992). Also for observations about Thomas Edison's achievements in the context of innovation, see Andrew Hargadon, *How Breakthroughs Happen: The Surprising Truth About How Companies Innovate* (Boston: Harvard Business School Press, 2003), and for a summary

at http://erc.atdc.org/documents/BreakthroughHappens.pdf. He discredits the idea of innovation being the result of a "lone genius."

CHAPTER 4: CREATING CUSTOMER VALUE: YOUR ONLY JOB

[1] Told to us by Sarah Nowlin in 2005 about her father, Bob Ginnings, former president of Hekimian Labs.

[2] Ellen Wills, "Why Professors Turn to Organized Labor," *New York Times,* May 29, 2001.

[3] Bill Bree, "The Hard Life and Restless Mind of America's Education Billionaire," *Fast Company,* March 2003.

[4] Donald B. Irwin and Beverly A. Drinnien, *Psychology—The Search for Understanding* (New York: West Publishing Company, 1987).

[5] We first heard of this chart more than twenty years ago and its origins are not known to us. "Quality" in this case is commensurate with fidelity or vividness, and "convenience" can also be thought of as control over the electronic device.

[6] Jonathan Mahler, *The Lexus Story* (New York: Melcher Media, 2004).

[7] Is it any wonder it is difficult to set the timer on your VCR? Here, from the Web, is a typical set of instructions:

1. Back of VCR set on 4.
2. Front of VCR set on channel you wish to record.
3. Turn TV to channel 4.
4. Put blank tape in VCR.
5. Set clock on VCR:
   a. Push T-ADJ button.
   b. Push DAY button to correct day.
   c. Push TIME button to correct time.
   d. Push T-ADJ button to stop flashing.
6. Set timer on VCR:
   a. Push P-CHECK button to turn on programming mode. You will be setting the 1st program.
   b. Set time for correct time to start recording.
   c. Set recording speed: SP(2hrs), LP(4hrs), EP(6hrs).
   d. Set channel for channel you wish to record.
   e. Push P-CHECK button again to set timer for when it is to turn off.
   f. Set time for when you want recording to stop. Speed and channel will remain the same.

g. Push P-CHECK again, 4 more times if you do not wish to set any more programs to be recorded on the timer.

7. TURN THE POWER TO THE VCR OFF!

    a. The front of the VCR will light up with the word TIMER. If it is flashing then you don't have a tape in the VCR.

    b. Please note this last step is crucial. If you don't turn the power to the VCR off, the timer won't work.

[8] From Michael Markowitz, business consultant, in a personal communication to Curtis R. Carlson in 2004.

[9] Great products often have several levels of value. One of our partners refers to the "three surprise rule" as a test for great products. These are products that will get customers talking to their friends about how terrific it is. Consider, for example, a "three surprise" cell phone. When you look at it and feel it, you are first pleasantly surprised by its striking design qualities. Then, when you use it as a basic telephone, you are again surprised because it is so convenient. Finally, when you use one of the more subtle features, like the video camera, you are again surprised by how intuitive it is to operate.

[10] The definition, Customer Value = Benefits − Costs, does not seem for some to represent "perceived" value as well as Value Factor = Benefits/Costs. Many human perceptual quantities are logarithmic, since nature had to create systems that could operate over many orders of magnitude. Consider human vision as an example. It operates over six orders of magnitude, from starlight to direct sunlight, and it responds logarithmically to brightness. "Value" can exist over many orders of magnitude too, from pennies to hundreds of thousands of dollars. Issues like health and security are even "priceless" to us. If we consider the simplest representation of these ideas and assume that quality, convenience, and their costs are perceptually independent, logarithmic quantities, then perceived value is propositional to log (Quality benefits) + log(Convenience benefits) − log(Quality costs) − log(Convenience costs) = log[(Quality benefits) × (Convenience benefits)]/[(Quality costs) × (Convenience costs)]. Finally we note that a computer's quality benefits are proportional to speed (bits per second). If we assume fixed convenience benefits and convenience costs, then a measure of perceived computer value is log (Speed/Quality costs). If, like for other perceptual quantities, people require a fixed change in this quantity to produce a noticeable improvement, then this observation would help explain the exponential behavior of Moore's Law. Moreover, we estimate that a just-noticeable difference in computer performance is roughly a 10 percent improvement in bits per second for a fixed cost. For a customer to notice a significant change, the improvement must be many times this value, say 100 percent. This is what is being achieved today every eighteen months. Clearly these issues and the assumptions behind them require additional study.

[11] Harry Cook has an unusual profession. Although he is a retired full professor of

engineering from the University of Illinois, he is really a marketer. Previously, he had been the head of research at Chrysler, where he became convinced that customer value and speed of innovation are the keys for thinking about the merits of a new product or service. As he rhetorically says, "What else could be more important?" When you talk to him, he seems frustrated that this is not obvious to everyone. In conversation, he skips immediately to the conclusion of his arguments, as if everyone understands what he is saying. Few do.

Harry Cook makes important observations that help quantify the value of a product or service. We will first start with a few more definitions. Value to the enterprise or seller is:

Seller value = Product price − Seller costs

In addition, call the product's total price P, the total benefits B, and the total costs to the seller C. With these definitions, consider the case where the customer and seller bargain equally with equal information. They should arrive at a price that makes the customer value and the seller value the same. Thus,

Customer value = (B − P) = Seller value = (P − C),

and thus,

B = 2 P − C,

which allows an estimation of the worth, B, of the benefits when the price and cost are known. Likewise, if customer value and cost are known for a product, the selling price can be estimated.

It can be shown that these conditions define an important case, the point of maximum cash flow (i.e., profit) for a monopoly. This is the price Microsoft, for example, should pick for its products. If Microsoft 2000 software has a price of $200 with a total cost of $100, then the worth of the product, B, is approximately $300, and the resulting customer value is $100.

What happens when there is competition? Another important case, which is predicted by another model, says that when there are five competitors making similar products,

P = B / 2.

Many businesses have roughly five competitors, like the automobile and mobile phone businesses. Thus, one can readily estimate the worth of a product's benefits, B, in those markets. Consider, for example, if Microsoft had five competitors and the product's worth was still $300; then the price Microsoft could charge would drop to $300/2, or $150. The price of $150 is a lot less than the monopoly price of $200, which shows the advantage of being a monopoly. The resulting customer value in this case would increase to $150.

As the number of competitors increases, the price that can be charged starts

approaching the cost to make the product. By the time there are ten or fifteen competitors, it is hard to make a reasonable profit. These "commodity" industries are tough places to make a living. Survival usually requires reducing competition through consolidation.

Cook uses these results and shows how to gain additional insights about customer value resulting from specific product attributes by using simple analytical models calibrated using preference studies. For example, Cook asks, "What is the potential customer value resulting from reduced car noise?" He notes that at one limit, where the noise is at the threshold of pain—110 dB—the customer value for the car goes to zero. At the other limit, experiments show that background noise should be greater than 40 dB because below that level it is "too quiet," which is a level that makes people feel uncomfortable. Cook fits these two end points with a simple curve and performs a series of calibration experiments, where he has potential consumers listen to two levels of noise assuming a luxury car. In each experiment he asks them, "How much more would you pay for the quieter car?" From conducting a series of experiments with different noise levels, he estimates that for a car costing $40,000, each dB reduction in car noise is worth approximately $400 in new customer benefits (remembering that B = 2P from above). This is a significant result that would warrant additional studies to determine its validity and whether it could be technically and financially achieved. See Harry E. Cook, *Design for Six Sigma as Strategic Experimentation: Planning, Designing, and Building World-Class Products and Services* (Milwaukee, WI: ASQ Quality Press, 2004); and Harry E. Cook, *Product Management: Value, Quality, Cost, Price, Profit and Organization* (New York: Kluwer Academic Publishers, 1997).

CHAPTER 5: IT'S AS SIMPLE AS NABC: HOW LIZ GOT HER BIG JOB

[1] See, for example, Mihaly Csikszentmihalyi, *Creativity: Flow and the Psychology of Discovery and Invention* (New York: Harper Perennial, 1997).

[2] Twyla Tharp, *The Creative Habit: Learn It and Use It for Life* (New York: Simon & Schuster, 2003).

CHAPTER 6: WATERING HOLES FOR CREATING VALUE: THE DAY THE BBC WALKED IN

[1] Although not the same, we were inspired by Edward Bono, *Six Thinking Hats* (Boston: Back Bay Books, 1999).

[2] Alex F. Osborn, *Applied Imagination: Principles and Procedures of Creative Thinking* (New York: Scribner, 1953).

[3] Adrian F. Furnham, "The Brainstorming Myth," *Business Strategy Review*, No. 4, 11,

2000, pp. 21–28. IDEO, the design firm in Palo Alto, California, also uses an extension of brainstorming when it insists on "observation, brainstorming, proto-typing, and implementing." Classic brainstorming is not enough.

4 *Ibid.*

5 For the current list of applications from Artificial Muscle, Inc., see http://www. artificialmuscle.com/applications/.

CHAPTER 7: MORE IDEAS FOR FASTER VALUE CREATION: ORIGINS OF LINUX

1 Douglas Engelbart is a fellow of the Computer History Museum in Mountain View, California. A version of this quotation can be found at http://www.computer history.org/fellows2005/bios/engelbart.shtml.

2 "Linux," Wikipedia, at http://en.wikipedia.org/wiki/Linux, provides an overview of Linux and the details of its development, including credit to more of the con-tributors. Much of the software used for the first version of Linux came from others. The key system software, including the C compiler, came from the Free Software Foundation's GNU project. The GNU project, started in 1984, had the goal of de-veloping a free Unix-like operating system.

3 The quality on Wikipedia has become an important topic of study. See Daniel Brandt at http://www.wikipedia-watch.org/hivemind.html for a critique. See, also, "Wikipedia Survives Research Test," BBC, December 15, 2005, at http://news. bbc.co.uk/2/hi/technology/4530930.stm, for a summary of tests done to measure the quality of Wikipedia's content. In general, it seems to compare favorably with commercial encyclopedias. "The free online resource Wikipedia is about as accurate on science as the Encyclopedia Britannica," but "[t]hey need a good editor." Changes have been proposed to address some of the more obvious limitations of Wikipedia.

4 Kenneth Klee, "Rewriting the Rules in R&D," *Corporate Dealmaker,* December 13, 2004, pp. 14–21. See also the Procter & Gamble "Connect + Develop" brochure distributed by P&G, 2003, and available on their website at http://www.pg connectdevelop.com.

5 Larry Huston, "Rewriting the Rules in Innovation," Business Innovation Factory, 2005, at http://www.businessinnovationfactory.com.

6 Find InnoCentive on the Web at http://www.innocentive.com.

7 See the *Wall Street Journal* at http://online.wsj.com.

8 "Sarnoff's Law," Wikipedia, at http://www.infoanarchy.org/wiki/index.php/Sarnoff's _Law. More information about David Sarnoff can be found in Eugene Lyons, *David Sarnoff: A Biography* (New York: Harper & Row, 1967).

9 George Gilder, *Telecosm: How Infinite Bandwidth Will Revolutionize Our World* (New York: Free Press, 2000). The number of network interconnections for Metcalf's

Law is equal to ($N^2 - N$) where N is the number of users. In the early days of the Internet, people proposed proprietary systems that would have split the Internet into isolated communities of users. Going between these different communities would have been impractical, and the value of the Internet would have been dramatically reduced. Metcalfe originally came up with his law to convince people to make communication systems, like the Internet, transparent to one other in order to realize this enormous potential. It was a major contribution.

[10] David P. Reed, "That Sneaky Exponential: Beyond Metcalfe's Law to the Power of Community Building," Reed's Locus at http://www.reed.com.

[11] The equation for the law of exponential interconnections and the number of interconnections between N users is $N \times (2^{(N-1)} - 1)$. Consider a simple example to show how the law of exponential interconnections works. Imagine that you own the only fax machine in the world, so $N = 1$. Since you have the only fax machine, there is no one to fax to and your fax machine has *no value* as a communication device. This is no surprise: You cannot send a fax to anyone. If you talk a friend into buying a fax machine, then $N = 2$, and you can send faxes back and forth to each other. Now your fax machine has a potential communication value of 2. If another person joins you, there are three people with six possible *direct connections* (i.e., the two connections between all three people) plus three *broadcast* connections to different groups (i.e., where each user sends the same message to both users). Your fax machine now has a potential communication value of 9. If you talk someone else into buying a fax machine, then $N = 4$, and there are now twenty-eight possible user connections composed of twelve direct connections and sixteen different group broadcast connections. The potential communication value of your fax machine is increasing exponentially with the number of people connected.

It should now be clear that if you have a communication device, like a fax machine, you have a powerful reason to talk all your friends into buying one too. Each person who buys one makes yours more valuable. And they have the same reason to encourage *their* friends and business partners to buy fax machines too. Everyone with a fax machine becomes a salesperson for fax machines. Community after community of new users adds to the buzz about fax machines. This is an example of "viral marketing," where we "infect" others and create a network because it is in our interest to do so. Remarkably, as more people buy fax machines, they become both cheaper to buy and more valuable to use.

The communication value of the Internet is, in this way, like the fax machine. Value grows as more users, user communities, organizations, and companies become connected. Not surprisingly, the Internet also grows through viral marketing. If you are connected to the Internet, it is in your interest to spread the word about its value. And, of course, each new user spreads the word to others, adding to the potential value represented by the law of exponential interconnections.

[12] This Alan Kay quote can be found at numerous locations on the Web, including Folklore at http://www.folklore.org.

13 IDEO uses this approach, as described in "The Deep Dive," *Nightline: ABC News.* Also see Tom Kelley, *The Art of Innovation* (New York: Currency, 2001).

14 This quotation is a paraphrase from Epictetus, who said, "The reason we have two ears and only one mouth, is that we may hear more and speak less." See http://www.quotationspage.com.

15 Yvon Chouinard, *Let My People Go Surfing: The Education of a Reluctant Businessman* (New York: Penguin, 2005), p 42.

16 Depending on the industry, rates of significant success are disparate and vary from 1 out of 3 to 1 out of 20. For an analysis of reasons for failure, whatever the rate, see Gary Diffendaffer, "Tips for Small-Business Owners and the Self Employed," *Denver Business Journal,* July, 23, 2004, at http://www.bizjournals.com/denver/stories/2004/07/26/smallb5.html.

17 Louise Robbins, *Louis Pasteur: And the Hidden World of Microbes* (Oxford: Oxford University Press, 2001).

18 George Sands Bryan, *Edison: The Man and His Works* (New York: A.A. Knopf, 1926).

CHAPTER 8: YOUR ELEVATOR PITCH: HOW HDTV BEGAN

1 Many derivatives of this quotation are variously ascribed to Winston Churchill and Mark Twain. See http://www.brainyquote.com. Also, as Mason Cooley said, "A moment of eloquence enthralls us. An hour's worth leaves us stupefied." See Mason Cooley, *City Aphorisms* (New York, 1985).

2 MUSE stands for Multiple Sub-Nyquist Sampling and Encoding. See NHK at http://www.nhk.or.jp/.

3 William Strunk and E. B. White, *The Elements of Style* (New York: Pearson Education, 1999).

4 From the movie *Amadeus,* 1984, to which Mozart replied, "But sire, there are neither more nor less than are required."

5 See Sequoia Capital at http://www.sequoiacap.com.

6 From the movie *Working Girl,* 1988, directed by Mike Nichols; with Harrison Ford, Sigourney Weaver, Melanie Griffith, Alec Baldwin, and Joan Cusack.

7 Songbird Medical Systems was acquired by P&G in 2003.

8 Steven J. Spear, "The Health Factory," *New York Times,* August 29, 2005.

9 Andrew Hargadon, *How Breakthroughs Happen: The Surprising Truth About How Companies Innovate* (Boston: Harvard Business School Press, 2003). "Many people still believe a better mousetrap is all it takes. But of the 2000+ mousetraps patented, only two

have sold well, and they were both designed in the 19th century. A good idea doesn't sell itself although most 'lone inventors' make the mistake of thinking it will."

[10] SWAG = "Scientific Wild Ass Guess."

CHAPTER 9: YOUR INNOVATION PLAN: FROM THE SKI SLOPE TO THE FIREHOUSE

[1] See PacketHop at http://www.packethop.com.

[2] Geoffrey A. Moore, *Crossing the Chasm: Marketing and Selling High-Tech Products to Mainstream Customers* (New York: Collins, 2002).

[3] A. Garcia of the SBA as quoted by Betsy Seanard and Lloyd J. Taylor, "Goldratt's Thinking Process Applied to the Problem of Business Failures," ASBE, Albuquerque, NM, 2004.

[4] For the latest estimates, see the Quarterly Venture Capital Report by Ernst and Young at http://www.ey.com/global/content.nsf/International/Home, or, specific to Silicon Valley, see Silicon Beat at http://www.siliconbeat.com.

[5] Hal Plotkin, "Free Money Here! Online Help for Finding Venture Capital: Which You May Not Want to Use," SFGate.com, November 23, 1998, at http://www.sfgate.com.

[6] Here is the recommended format for Sequoia Capital at http://www.sequoiacap.com:

1. Business
   a. Company's business (description short enough to fit on a business card)
   b. Mission statement
2. Products
   a. Product description
   b. Development schedule
   c. Differentiation
   d. Price point
3. Market
   a. Trends
   b. Historic and projected sizes in dollars
   c. Product match to market definition
4. Distribution
   a. Sales channels
   b. Partnerships
   c. Customers

5. Competition

    a. Competitors

    b. Competitive advantages

6. Team

    a. Background of management

    b. Board composition

7. Financials

    a. Historic and projected P&L (first two years by quarters)

    b. Projected cash flow (first two years by quarters)

    c. Current balance sheet

    d. Projected head count by functional area (R&D, sales, marketing, G&A)

    e. Capitalization schedule

8. Deal

    a. Amount raised

    b. Valuation asked

    c. Use of proceeds

[7] A *vision statement* describes where you want to be in the future. It describes a state of accomplishment when successful. Vision statements apply to all projects. A *mission statement* gives the purpose and basis for being in business. New ventures require one, and established organizations already have one. They serve as guides to direct actions. Here are simple examples from Toyota Industrial Equipment: "Our Vision Statement guides our principles and business practices: To become the most successful and respected lift truck company in the United States. To reach our vision, we use our Mission Statement as a daily guide: To sustain profitable growth by providing the best customer experience and dealer support."

[8] "A beachhead is a military term used to describe the line created when a unit (by sea) reaches a beach and begins to defend that area of beach, while other reinforcements (hopefully) help out, until a unit large enough to begin advancing has arrived." In business it means getting a foothold in a small market from which you can then leverage to expand into a larger market. See Wikipedia at http://www.wikipedia.org.

[9] For commercial business opportunities, Satyam Cherukuri, president and CEO of the Sarnoff Corporation, emphasizes describing the Space, Positioning, Model, and Investment (SPMI). A compelling business plan is a comprehensive approach that outlines an opportunity in an ecosystem (Space) and proposes the optimal point of attack (Positioning) to generate significant revenues (Model) with minimal financial risk (Investment).

[10] One study suggests the average working family saves $2300 per year by shopping at Wal-Mart. "Measuring the Economic Impact of Wal-Mart on the U.S. Economy,"

Global Insight, at http://www.globalinsight.com/MultiClientStudy/MultiClient StudyDetail2438.htm.

[11] See ThinkExist.com at http://en.thinkexist.com.

[12] See Worcester Polytechnic Institute at http://www.wpi.edu.

[13] See Artificial Muscle at www.artificialmuscle.com.

[14] Richard Foster and Sarah Kaplan, *Creative Destruction: Why Companies That Are Built to Last Under Perform the Market—And How to Successfully Transform Them* (New York: Currency Publishers, 2001).

[15] Langdon Morris, "Business Model Warfare," Innovation Labs White Paper, Ackoff Center for the Advancement of Systems Approaches (A-CASA), the University of Pennsylvania, 2003.

[16] Peter Drucker, *Innovation and Entrepreneurship* (New York: Harper Business, 1985), p. 111.

[17] See Häagen-Dazs at http://www.haagendazs.com.

[18] There are many reasons why digital HDTV video looks better than 35mm film. One reason is that when 35mm film is projected, the film jitters as it is pulled through the film gate. This film gate jitter significantly reduces the actual resolution seen by moviegoers. Digital video does not suffer from this problem.

[19] "Film Studios Hunt Down Web Pirates," CBS News, November 16, 2004, at http://www.cbsnews.com. "The MPAA claims the U.S. movie industry loses more than $3 billion annually in potential global revenue because of physical piracy, or bogus copies of videos and DVDs of its films."

[20] Morris, *op. cit.*

CHAPTER 10: A CHAMPION: THE MAYOR OF KELLYVILLE

[1] " 'You've Got to Find What You Love,' Jobs Says," Stanford Report, June 14, 2005, at http://news-service.stanford.edu.

[2] Champions are:

- Builders
- Passionate, committed, and curious
- From all disciplines, all levels
- Synthesizers—focused on a vision
- Team and partnership creators
- Helpers who also continuously seek help from others
- Organizationally responsible

[3] J. C. R. Licklider, "Man-Computer Symbiosis," *IRE Transactions on Human Factors in Electronics,* Vol. HFE-1, March 1960, pp. 4-11.

4 This statement can be found in many sources, such as "Why Social Software Makes for Poor Recommendations" Usernomics, May 08, 2005 at http://www.usernomics.com. In 2003 Bill Joy left Sun Microsystems.

5 "Apple Computer," Wikipedia at http://en.wikipedia.org/wiki/Apple_computer.

6 "Hewlett-Packard," Wikipedia at http://en.wikipedia.org/wiki/Hewlett-Packard_Company.

7 SWAG = "Scientific Wild Ass Guess."

8 We emphasize the concept of champions and not leadership, management, or entrepreneurship. In traditional management books, a great deal of importance is given to leadership with examples of superb leaders, such as Jack Welch. Obviously, leadership is to be encouraged and celebrated. And clearly champions are leaders. But *leadership* is a term that can create confusion and misalignment within a team. The word often suggests a hierarchical organization: "I am in charge— follow me." In the exponential economy, we take turns being the leader. Each of us needs to be an empowered champion for our part of the project. Even if you are the CEO of a company, you will be subordinate to others on your team most of the time.

Telling someone to be a leader also causes uneasiness in many people. When we think of leadership, images like Churchill and Gandhi come to mind. How can anyone live up to those examples? But all of us have had the experience in our lives of being a passionate champion for some activity. It might have been teaching soccer, playing music, being on a sports team, or pursuing a childhood hobby. That feeling is something we can relate to, build on, and extend into a new opportunity or project. These previous experiences also illustrate the rewards of becoming a champion—working with others, learning new skills, achievement, and having fun.

What about the term *management*? If you are a manager, you may have been taught to focus within the company to get the job done, rather than outward to create maximum customer value. Most managers still fill the traditional "boss" role, which is an industrial-age concept. It evokes feelings of hierarchy, "we-they," and "control"—concepts inappropriate for the exponential economy. It won't attract and motivate the best people.

Management skills are, of course, critical for success on any project. Peter Drucker, the most important observer and writer about management as a profession, has eloquently described the critical roles, functions, and skills needed for good management. Those are even more important in the exponential economy. Clearly, champions are managers too. Someone must have overall responsibility within an organization and perform the three basic functions of management: pick the team, monitor the team, and reward the team. Our purpose is to recast those functions and others into a more positive, proactive, and inclusive framework.

Finally, we avoid using the word *entrepreneur,* although champions clearly have entrepreneurial attributes. That is another powerful and admirable concept, but it comes with considerable baggage, especially in Silicon Valley. An entrepreneur is sometimes thought of as a risk taker because so few start-up companies make it past five years. Some start-ups either fail completely or join the "living dead" that neither grow nor die. But the one thing successful entrepreneurs are *not* is risk takers. They *are* risk reducers. Like Thomas Edison, they do their homework, minimize risk, and succeed. We encourage people to do their homework and create the conditions necessary for success.

9 Johann Wolfgang von Goethe (translated by John Anster), *Faust* (London: Cassell, 1835), p. 20. There seems to be some question about the authenticity of this quote. Another famous quotation, which is also often attributed to Goethe, is:

> Until one is committed, there is hesitancy, the chance to draw back, always ineffectiveness.
> Concerning acts of initiative and creation there is one elementary truth, the ignorance of which kills countless ideas and splendid plans.
> That the moment one definitely commits then Providence moves too.
> All sorts of things occur to help one that would never have otherwise occurred.
> A whole stream of events issues from the decision, raising in one's favor all manner of unforeseen incidents and meetings and material assistance, which no person could have dreamt would come that way.
> Whatever you can do, or dream you can, begin it. Begin it now.

But this quotation is apparently by William Hutchinson Murray, *The Scottish Himalayan Expedition* (London: J. M. Dent & Sons, 1951). It was inspired by the quotation we gave from Goethe.

CHAPTER 11: GENIUS OF TEAMS: DOUGLAS ENGELBART AND THE BIRTH OF THE PERSONAL COMPUTER

1 While this exact quotation does not appear to be in print, this explanation from the Institute for Intercultural Studies is helpful: "Although the Institute has received many inquiries about this famous admonition by Margaret Mead, we have been unable to locate when and where it was first cited, becoming a motto for many organizations and movements. We believe it probably came into circulation through a newspaper report of something said spontaneously and informally. We know, however, that it was firmly rooted in her professional work and that it reflected a conviction that she expressed often, in different contexts and phrasings." "Frequently Asked Questions About Margaret Mead," Institute for Intercultural Studies, at http://womenshistory.about.com.

2 As Engelbart said, "I looked up and everyone was standing, cheering like crazy." "The Click Heard Around the World," *Wired* magazine, January 2004, at http://www.wired.com/wired/archive/12.01/mouse.html. Video clips of the presentation are available at MouseSite at http://sloan.stanford.edu/MouseSite/1968Demo.html. See the proceedings of a symposium held at Stanford University in 2000, "Engelbart's

Unfinished Revolution," at http://unrev.stanford.edu/. Engelbart's Bootstrap Institute can be found at http://bootstrap.org, which includes additional presentations given at Stanford University. "The Unfinished Revolution: Strategy and Means for Coping with Complex Problems," April 2000.

3 Amir D. Aczel, *Fermat's Last Theorem: Unlocking the Secret of an Ancient Mathematical Problem* (New York: Delta, 1997). Fermat's Last Theorem, $Z^n = X^n + Y^n$, has no non-zero integer solutions for X, Y, and Z when $n>2$. As equations go, this one doesn't look *that* hard. Fermat made it more intriguing by writing on the margin of his notebook, "I have discovered a truly remarkable proof, which this margin is too small to contain."

4 For a detailed discussion of IDEO's recursive processes, see Tom Kelley, *The Art of Innovation* (New York: Currency Publishers, 2001).

5 Eventually, these "agents" may be smarter than we are. Ray Kurzweil, *The Singularity Is Near: When Human Transcend Biology* (New York: Viking, 2005), p. 136. It is interesting to contemplate the rate of growth in certain fields in the exponential economy when billions of computer agents, thousands of times smarter than humans, are interacting at the speed of light. A NIC working to solve the world's hardest problems could include thousands of other related NICs without difficulty.

6 Tapped In can be located at http://tappedin.org/tappedin/.

7 In music, it is accepted that to learn a new piece one must repeat it dozens of times. To unlearn a bad habit takes two to ten times as much effort. When learning new ideas, teams are similar.

CHAPTER 12: FORMING THE INNOVATION TEAM: HOW WE WON AN EMMY FOR HDTV

1 In addition to the Sarnoff Corporation, the four other companies in the Advanced Television Research Consortium (ATRC) were Thomson, Philips, NBC, and CLI.

2 We asked Norm why he did not shorten anyone's plans. He said, "They had already squeezed as much out of their parts as they could. I could see in their faces how nervous they were and after talking to them, I decided whether to leave their part alone or to increase the time they had."

3 Tracy Kidder, *The Soul of a New Machine* (New York: Back Bay Books, 2000).

4 Picture thanks to Glenn Reitmeier.

5 There were three parts to the HDTV system: an encoder, a transmitter, and the decoder. When the encoder and decoder were connected, the system worked. When the transmitter was connected, it did not. The problem turned out to be that the connecting wires for the encoder and decoder systems were wired the same, but incorrect, way. The transmitter was wired correctly. Finally, a teammate checked the connectors and found the mistake.

[6] Notably there were Jim Carnes, Mike Isnardi, Charlie Wine, and all our partners from Thomson, including Eric Geiger and Jacques Sabatier, plus the teams from Philips, NBC, and CLI. After the developments described here, the saga was far from over. In 1993, the FCC encouraged the competing systems to combine into a single "best-of-the-best" system. Again, former competitors would become allies, and the so-called "Grand Alliance" system did indeed establish the U.S. HDTV standard, including many of the key parts of the initial Sarnoff prototype system. For the second phase of the development, the Grand Alliance system, the partner companies included Sarnoff, AT&T, General Instrument, MIT, North American Philips, Thomson, and Zenith. See "The Path of United States HDTV World: A Brief on United States HDTV," Ecoustics.com, November 11, 2004, at http://forum.ecoustics.com/bbs/messages/34579/108629.html.

[7] See Barnum Associates International at http://www.barnumassociates.com.

[8] The Dilbert website provides randomly generated vision statements, although we didn't use it to generate the one shown. See Dilbert.com at http://www.dilbert.com.

[9] See the Collaboration Institute at http://www.CollaborationInstitute.com.

[10] See PatchWorx at http://www.patchworx.org. Theresa Middleton started this site "to provide a safe and secure online community supporting children facing serious illness or disability."

[11] Jim Collins, *Good to Great: Why Some Companies Make the Leap and Others Don't* (New York: HarperCollins, 2001).

[12] William W. Wilmot, *Relational Communication* (New York: McGraw-Hill, 1995).

[13] Frederick Phillips Brooks, *The Mythical Man-Month: Essays on Software Engineering* (Boston: Addison-Wesley, 1975).

[14] Conflict management is an extremely important skill. For a primer on conflict management at work, see William W. Wilmot and Joyce L. Hocker, *Interpersonal Conflict*, 7th edition (New York: McGraw-Hill, 2007); and for intervention skills, see Elaine Yarbrough and William Wilmot, *Artful Mediation: Constructive Conflict at Work* (Boulder, CO: Cairns Publishing, 1995).

CHAPTER 13: OVERCOMING BLOCKAGES TO INNOVATION: JIM TORPEDOES A SPLENDID IDEA

[1] Brent Schlender and Christene Y. Chen, "Jobs' Apple Gets Way Cooler," *Fortune*, January 2000.

[2] As quoted in Yvon Chouinard, *Let My People Go Surfing: The Education of a Reluctant Businessman* (New York: Penguin Books, 2005), p. 97.

[3] FUD was defined by Gene Amdahl after founding Amdahl Corp. He said, "FUD is the fear, uncertainty, and doubt that IBM sales people instill in the minds of potential customers who might be considering Amdahl products." For the application of

FUD to different fields, consult "Fear, Uncertainty, and Doubt," Wikipedia at http://en.wikipedia.org/wiki/FUD.

4 For a list of the specific skills mediators use in the workplace, consult Elaine Yarbrough and William Wilmot, *Artful Mediation: Constructive Conflict at Work* (Boulder, CO: Cairns Publishing, 1995).

5 Emily Litella was an old lady television character on NBC's *Saturday Night Live* who would launch into tirades on various topics, always based on a false premise. When the mistake was revealed, Emily would simply look into the camera and quietly say, "Never mind." See "Gilda Radner," Wikipedia, at http://en.wikipedia. org/wiki/Gilda_Radner.

6 Yarbrough and Wilmot, *op. cit.*

CHAPTER 14: INNOVATION MOTIVATORS: SAVING LARRY'S LIFE

1 For other Eleanor Roosevelt quotes, go to Brainy Quote at http://www.brainy quote.com.

2 Peter Drucker, *The Practice of Management* (New York: Collins, 1993); Peter Drucker, *Management: Tasks, Responsibilities, and Practices* (New York: Harper Business, 1993). See also Drucker's publications at the Leader to Leader Institute that he started for nonprofit organizations at http://www.leadertoleader.org.

3 For a discussion of how these inclusion issues play out in conflicts by manifesting as identity and relationship issues, see Chapter 3 of William W. Wilmot and Joyce L. Hocker, *Interpersonal Conflict,* 7th edition (New York: McGraw-Hill, 2007).

4 As OSHA (Occupational Safety and Health Administration) writes, "In a strong safety culture, everyone feels responsible for safety and pursues it on a daily basis; employees go beyond 'the call of duty' to identify unsafe conditions and behaviors, and intervene to correct them." See the U.S. Department of Labor at http://www.osha.gov.

5 Philip H. Mirvis and Mitchell L. Marks, *Managing the Merger: Making It Work* (New York: Beard Books, 2003). Most estimates say that between 50 percent and 85 percent of mergers and acquisitions fail, usually because of not involving people who must execute these significant changes. See the CEO Refresher at http://www. refresher.com/!laswhy.html.

6 Jennifer Pittman, "Why Do Employees Leave? Study Says It's Not for Money," *Business Journal,* January 4, 2002, at http://www.bizjournals.com. Another study of newspaper employees who left wrote, "Respondents listed the following issues, in decreasing order of importance: fairness in promotions; involvement in decision-making; opportunities for advancement; supervisor concern for employees' personal success; fairness in pay; equitable treatment; and contributions in value."

Newspaper Association of America and Newspaper Personnel Relations Association study, *Editor & Publisher,* vol. 128, no.18, May 6, 1995, p. 23(1).

[7] One of the most widely cited scholarly works on motivation is Frederick Hertzberg, "One More Time: How Do You Motivate Employees," *Harvard Business Review,* January and February 1968, pp. 87–96. His list of motivators starts with:

1 = Achievement
2 = Recognition
3 = Work itself
4 = Advancement
5 = Growth

Salary is only 11 on his list and it is one of the "hygienic factors." That is, you cannot motivate people with it—you can only demotivate people if you get it wrong. This list is an excellent summary of the benefits that come from being a champion and part of a high-impact innovation team.

CHAPTER 15: YOUR INNOVATION TEAM: YOU CAN START NOW

[1] William Hutchinson Murray, *The Scottish Himalayan Expedition* (London: J. M. Dent & Sons, 1951).

[2] Satyam Cherukuri lists three phases of United States research and development in this century. The first was based on Edison's concept of the corporate research laboratory. The second was based on a distributed venture model. The third, which we are now part of, is the global model, where innovation resources are distributed throughout the world.

[3] Taiichi Ohno, *Toyota Production System: Beyond Large-Scale Production* (New York: Productivity Press, 1988). As with any seminal concept, however, others give a different origin to "Just in Time." See Neil A. Grauer at The Gazette Online, the Newspaper of the Johns Hopkins University, at http://www.jhu.edu/~gazette/1999/oct2599/25ford.html. He notes that it might be an offshoot from Henry Ford's work.

[4] Ohno, *op. cit.* Toyota uses three terms for activities that detract from value creation. They translate as MURI = overburden, MUDA = waste, and MURA = under-utilization. For ease of use, we use the MUDA term for all non-value-adding activities.

CHAPTER 16: THE INNOVATION ENTERPRISE: CONTINUOUS VALUE CREATION (CVC) THROUGHOUT

[1] Preface to Toyota handbook, *Benchmark, Toyota Production System,* no date. Also see *Benchmark Toyota Basic Management Philosophy,* no date.

[2] Gary Hamel, *Leading the Revolution* (New York: Penguin Group, 2002).

3 Andrew C. Inkpen, "Learning Through Alliances: General Motors and NUMMI," available online product #CMR320, 24 pp., published August 01, 2005, Harvard Business Online, http://harvardbusinessonline.hbsp.harvard.edu.

4 Paul S. Adler, "The Learning Bureaucracy: New United Manufacturing, Inc.," in Barry M. Staw and Larry L. Cummings (eds.), *Research in Organizational Behavior,* Vol. 15 (Greenwich, CT: JAI Press, 1993), pp. 111–94. This is an excellent review of the successes and problems at NUMMI.

5 Andrew C. Inkpen, "Creating Knowledge Through Collaboration," available online product #CMR070, 20 pp., published October 1, 1996, Harvard Business Online.

6 At its peak, Ford was producing a Model T every 24 seconds. See http://inventors. about.com/library/inventors/blford.htm.

7 For cogent descriptions of Ohno's contributions, see econpapers.repec.org/paper/ tkyjseres/97j04.htm and http://curiouscat.com/guides/ohnobio.cfm.

8 Employee turnover at Ford approached 400 percent per year by 1913! See Robert Pollin and Stephanie Luce, *The Living Wage: Building a Fair Economy* (New York: New Press, 2000); http://encarta.msn.com/encyclopedia_761563934_3/ Automobile_Industry.html#p38. A cogent summary of the history of Henry Ford and his impact is at www.willamette.edu/~fthompso/MgmtCon/Henry_Ford.html. "People simply cannot be treated like robots. They have needs that robots don't have, and they have emotions that robots don't have. Essentially, trying to depersonalize workers in the workplace was doomed to failure." Interview with John J. Jarvis, director of ISyE program at Georgia Tech, found at http://gtalumni.org/ Publications/magazine/win95/isyechal.html.

9 Sharp Corporation, for example, has this as its business philosophy:

> We do not seek merely to expand our business volume.
> Rather, we are dedicated to the use of our unique,
> innovative technology to contribute to the culture,
> benefits and welfare of people throughout the world.
>
> It is the intention of our corporation to grow
> hand-in-hand with our employees,
> encouraging and aiding them to reach their full potential
> and improve their standard of living.
>
> Our future prosperity is directly linked to the prosperity of
> our customers, dealers and shareholders . . .
> indeed, the entire Sharp family.

See http://sharp-world.com/corporate/info/bp/index.html.

10 "*Nemawashi* in Japanese company culture is an informal process of quietly laying the foundation for some proposed change or project, by talking to the people con-

cerned, gathering support and feedback, and so forth. It is considered an important element in any major change, before any formal steps are taken, and successful *nemawashi* enables changes to be carried with the consent of all sides. *Nemawashi* literally translates as 'going around the roots.' " See http://en.wikipedia.org/wiki/ Nemawashi. As Tony Newman says, "The concept of *nemawashi* refers to the idea that when you want to dig a tree out, you don't just go and dig around one day and chop it down. You go out and very gently dig around the roots of the tree and then rock it a little bit, then you go back a week later and you dig a bit more and rock it a little bit, till finally on the day that you want to dig the tree out you just lean on it and it rolls over. In my experience you don't confront the Japanese. You work with them and you slowly move them in your direction. Confrontation and raised voices aren't something that I think works in Japanese culture."

[11] This quotation is attributed to many. One version is "If not now, when; if not us, who?" But the original comes from the Talmud.

[12] The "STEP" model from Lisa Friedman and Herman Gyr, *The Dynamic Enterprise* (San Francisco: Jossey-Bass, 1997), is one helpful approach to discussing issues of organizational alignment. It highlights (1) *External Environment.* The external environment includes marketplace opportunities, competition, suppliers, economy, regulations, resources, and potential partners. This is your market space and ecosystem; (2) *Task.* The products and services you create, manufacture, and deliver to your customers; (3) *Structure.* The formal organization, job descriptions, physical facilities, information systems, management policies, and incentive systems that make the task possible; (4) *People.* The skills and talents of people, how they communicate with one another, and the quality of their work relationships; and (5) *Internal Environment.* The values, attitudes, subconscious beliefs, and overall culture of the organization.

[13] This quotation is attributed variously to David Hanna, Steven Covey, A. Jones, J. Bockerstette, T. Kight, and an "old truism."

[14] Jim Collins, *Good to Great: Why Some Companies Make the Leap and Others Don't* (New York: HarperCollins, 2001).

[15] For a more complete picture of the organizational alignment actions needed, see Friedman and Gyr, *The Dynamic Enterprise.*

CHAPTER 17: INNOVATION'S FIVE DISCIPLINES: A FOUNDATION FOR NATIONAL COMPETITIVENESS IN A WORLD OF ABUNDANCE

[1] Jeff Colvin, "America: The 97 Lb. Weakling," *Fortune,* July 25, 2005. He quotes Chambers as saying, "We are not competitive," p. 5 at http://www.mckinsey. com/aboutus/mckinseynews/pressarchive/pdf/FortuneMcKstory.pdf. Note that we included "U.S." as the referent in the quotation so that it is clear he is talking about the United States. Other assessments by Chambers are included in Randy

Barrett, "Technology CEOs Cite Top Threats To U.S. Competitiveness," Technology Daily PM, March 8, 2005, at http://www.technet.org/tnd_news/techdaily_030805/. For one of John Chambers's latest assessments, see John Chambers, "Perspective: Answering Bush's Competition Challenge." CNET, February 1, 2006, at http://cnetnews.com. Consult also David Kirkpatrick, "Cisco CEO on U.S. Education: 'We're Losing the Battle,' " *Fortune,* March 30, 2005. The National Council on Competitiveness, in their survey of competitiveness in 2005, notes, in referring to their 2005 National Innovation Survey, that "American business executives are enthusiastic about innovation prospects globally, according to a new survey released today by the Council on Competitiveness, a finding that is consistent with the growing evidence of significant investments abroad in science and technology. On the home front, the survey found that executives are neutral to slightly positive about the innovation climate in the Untied States. 'This survey should serve as a wake-up call to business leaders across the United States,' said Deborah L. Wince-Smith, president of the Council on Competitiveness. "Innovation is perhaps the single most important factor in maintaining the U.S. competitive advantage in global markets and in driving a higher quality of life in the long run and it must be integrated at all levels into our corporate cultures." Retrieve the full report at the Council on Competitiveness at http://www.compete.org.

2   For current figures on venture capital, see the National Venture Capital Association, http://www.nvca.org/, the Dow Jones reports at http://www.dowjones.com, and venture capital reports from such sites as Research and Markets at http://www.researchandmarkets.com/info/full.asp.

3   For ongoing reports, consult PricewaterhouseCoopers at http://www.pwcmoney tree.com.

4   J. Srinivasan, "Will King Abdullah Keep Saudi Economy Well-Oiled?," Business Line at http://www.thehindubusinessline.com. "While the oil boom of the 1970s fetched Saudi Arabia billions of dollars to build roads, airports, and telecom systems that pushed a reluctant people into the 20th century, the worry now is that a stagnating economy is rapidly falling behind in the entrepreneurial- and technology-driven new millennium. For, much of Saudi Arabia's industry is state-controlled monopolies. Useful when developing the kingdom, many of these organizations have become liabilities now, stifling competition and discouraging investments. Saudi Arabia's low level of investment—16.7 percent of GDP against an average of 27 percent for all developing countries—is a main reason for the kingdom's disappointing economic performance, a Business Week report quotes the chief economist at National Commercial Bank in Jeddah. Further, unemployment is rising, and the Saudi economy is nowhere near creating enough jobs. Joblessness, according to the Bank, may be 14 percent."

5   Michael Porter and Scott Stern, "The New Challenge to America's Prosperity: Findings from the Innovation Index," Council on Competitiveness, Washington, D.C., 1999, p.36. The competitiveness index, computed by R&D expendi-

tures, growth in scientific and technological personnel, and spending on education, was computed in 1999 and predicted a drop for 2005.

6 "2005 National Innovation Survey, Executive Summary," Council on Competitiveness, Washington, D.C., 2005.

7 "National Innovation Initiative Summit and Report," Council on Competitiveness, Washington, D.C., 2005, p. 38. For the full citations for these statistics, consult the footnotes in the report.

8 Colvin, *op. cit.* See also TechNet at http://www.technet.org.

9 Consult the Program for International Assessment at http://nces.ed.gov/surveys/pisa.

10 Robert E. Slavin, "Evidence-Based Reform: Advancing the Education of Students at Risk," Center for American Progress at http://www.americanprogress.org; and Robert E. Slavin, "Renewing Our Schools, Securing Our Future," Center for American Progress, at http://www.ourfuture.org. Since rankings and estimates change frequently, consult these websites.

11 "Program for International Assessment," part of the National Center for Educational Statistics at http://nces.ed.gov/surveys/pisa.

12 Peter F. Drucker, *Forbes,* March 1997, p. 10; Krysten Crawford, "A Degree of Respect for Online MBAs," *Business 2.0,* December 2005, p. 102. Online education is having exponential growth rates. Enrollments in online MBA programs have gone from essentially zero in 1990 to 125,000 in 2005 with more than 150 accredited business schools.

13 For information on GDP and a variety of related statistics, consult the latest information at Bureau of Labor Statistics at http://www.bls.gov.

14 "California Schools Lag Behind Other States on Almost Every Objective Measure," Rand Corporation, January 3, 2005, at http://www.rand.org.

15 Ranking for universities are available at the Institute of Higher Education at http://ed.sjtu.edu.

16 Hennessy also said, "We have relied on imported labor. That situation has gotten more difficult after Sept. 11. It's going to be made more difficult by the fact that it's now attractive to go back to India, Taiwan, Korea or mainland China after you do your graduate work in the United States. When I first came to Stanford, nobody went back. Now, you see people go back right after school. That's a real change, and it's going to affect our ability to have the talent we need to make this country successful." "On the Record: John Hennessy," *San Francisco Chronicle,* October 19, 2003.

17 See the Taulbee Survey at http://www.cra.org/statistics.

18 Dominic Basulto, "An Innovation Roadmap," TCS Daily, December 19, 2005, at http://www.tcsdaily.com/article.aspx?id=121605F.

[19] "Nanotechnology in China Is Focusing on Innovations and New Products," Physorg.com, August 17, 2005, at http://www.physorg.com/news5870.html.

[20] "National Innovation Initiative Summit and Report," Council on Competitiveness, Washington, D.C., 2005, p. 38. For the full citations for these statistics, consult the footnotes in the report.

[21] Patrick Thibodeau, "Update: H-1B Visa Cap Reached; IT Groups May Press for More," Computerworld, August 12, 2005, at http://www.computerworld.com/careertopics/careers/story/0,10801,103883,00.html.

[22] Colvin, *op. cit.*

[23] See Pierrette Hondagneu-Sotelo, "Gendering Migration," at http://cmd.princeton.edu/papers/wp0502f.pdf. For the latest figures, consult the U.S. Citizenship and Immigration Services at http://uscis.gov.

[24] "While in the past, immigrants filled important job niches in the US economy, the situation is changing. In Congress, Rep. Lamar Smith remarked that immigrants are increasingly unsuited to the U.S. workforce." "First: Immigrants will account for half of the increase in the workforce in the 1990s. Second: The skill level of immigrants relative to Americans has been declining for years—35 percent of immigrant workers who have arrived since 1990 do not have a high school education, compared with 9 percent of native-born workers. Some 300,000 legal immigrants without high school educations arrive each year—and will total three million this decade. Third: Close to 90 percent of all future jobs will require post–high school education." Quoted in *Alamance Independent* at http://www.alamanceind.com/newfol~4/immig_7.html. The Center for Immigration Studies's "Current Numbers" at http://www.cis.org/topics/currentnumbers.html, gives this historical overview: "During the 1990s, an average of more than 1.3 million immigrants—legal and illegal—settled in the United States each year. Between January 2000 and March 2002, 3.3 million additional immigrants have arrived. In less than 50 years, the U.S. Census Bureau projects that immigration will cause the population of the United States to increase from its present 288 million to more than 400 million. The foreign-born population of the United States is currently 33.1 million, equal to 11.5 percent of the U.S. population. Of this total, the Census Bureau estimates 8–9 million are illegal immigrants. Other estimates indicate a considerably higher number of illegal immigrants. Approximately 1 million people receive permanent residency annually. In addition, the Census Bureau estimates a net increase of 500,000 illegal immigrants annually. The present level of immigration is significantly higher than the average historical level of immigration. This flow may be attributed, in part, to the extraordinary broadening of U.S. immigration policy in 1965. Since 1970, more than 30 million legal and illegal immigrants have settled in the United States, representing more than one-third of all people ever to come to America's shores. At the peak of the Great Wave of immigration in 1910, the number of immigrants living in the U.S. was less than half of what it is today, though the percentage of the population was slightly higher. The annual arrival of 1.5 mil-

lion legal and illegal immigrants, coupled with 750,000 annual births to immigrant women, is the determinate factor—or three-fourths—of all U.S. population growth."

25 See National Science Foundation, "National Patterns of R&D Resources," http://www.nsf.gov.

26 Consult "NSF Budget Falls in 2005," AAAS, at http://www.aaas.org/spp/rd/nsf05c.htm.

27 "Summary of Sarbanes-Oxley Act of 2002," AICPA, at http://www.aicpa.org/info/sarbanes_oxley_summary.htm.

28 For updated numbers, consult the U.S. Small Business Administration at http://www.sba.gov.

29 For estimates of the costs of SOX compliance, see "Sarbanes-Oxley: A Price Worth Paying?," Economist.com, May 19, 2005, at http://www.economist.com/business/displayStory.cfm?story_id=3984019; and also John Berlau, "A Tremendously Costly Law," *National Review,* April 11, 2005, at http://www.kauffman.org/pdf/tt/Berlau_John.pdf.

30 There is some encouraging news. Groups are beginning to suggest changes in the Sarbanes-Oxley requirements, among other measures. See, for example, Bob Greifeld, "It's Time to Pull Up Our SOX," *Wall Street Journal,* March 6, 2006. Other positive developments include: the United States House Democrats on November 15, 2005, released "The Innovation Agenda: A Commitment to Keep America #1," which, among other things, suggests that Congress should "[p]rovide small businesses with the tools to encourage entrepreneurial innovation and job creation." See http://www.housedemocrats.gov/; and President Bush, in his 2006 State of the Union address, called for a national initiative in science and math. See http://www.whitehouse.gov.

31 "Under Schroeder, Germany suffered its longest postwar economic slump, with growth of less than 1.3 percent in three of the past four years. The economy stagnated in the second quarter of this year. The unemployment rate rose to 12 percent in March, a postwar record." "Merkel, Schroeder Both Claim Right to Lead Germany," Bloomberg.com, September 18, 2005.

32 Neal E. Boudette, "German Labor Laws Zap Company's Profits," *Wall Street Journal* Online at http://www.careerjournaleurope.com.

33 Thomas Friedman, "Follow the 'Leapin' Leprechaun," *New York Times,* July 1, 2005.

34 "Wearing of the Green: Irish Subsidiary Lets Microsoft Slash Taxes in U.S. and Europe," *Wall Street Journal,* November 7, 2005, pp.1–10.

35 Jim Pinto, "The China Manufacturing Syndrome," *Automation World,* January 2005, and at http://www.jimpinto.com/writings/chinasyndrome.html.

36 Ho John Lee, "China Is Run by Engineers (Really)," at http://www.hojohnlee.com/weblog/archives/2005/06/08/china-is-run-by-engineers/. See also Jean Kumagai

and Marlowe Hood, "China's Tech Revolution," *IEEE Spectrum,* June 2005, at http://www.spectrum.ieee.org/jun05/1231.

[37] Richard K. Lester and Michael J. Piore, "National Innovation Summit Only Gets It Half Right," *Industry Week,* February 16, 2005.

[38] Michael Porter, "Clusters and the New Economy," *Harvard Business Review,* November/December, 1998.

[39] Azeezaly S. Jaffer, vice president, USPS Public Affairs and Communications, "Letter to the Editor," *Boston Herald,* 2005. "I read with interest your recent editorial 'Role Model for the USPS' (6/19/05) wherein you suggest that the Postal Service be privatized. I'm sure you're aware this is a public policy issue and it would have to be considered and decided by the Congress." "(A)s you note, the Postal Service is trying 'to be efficient.' I'll say it is! The Postal Service concluded fiscal year 2004 and 2003 comfortably in the black. Our expenses have remained relatively constant since 2001 and we've not increased prices since 2002. In addition, our on-time delivery scores and customer satisfaction index are at all time highs, as measured independently by IBM Consulting Services and the Gallup organization, respectively."

[40] Amelia Gruber, "IRS Employees Win Competitive Sourcing Study, Cut Jobs," National Contract Management Association, August 12, 2005, at http://www.govexec.com/story_page.cfm?articleid=31995&sid=6.

[41] See, for example, "Documents Available to DoD Users in the ISO 9000 & ISO 14000 Collections," Defense Standardization Program Office at http://www.dsp.dla.mil/ISO-list.htm.

[42] "National Innovation Initiative Summit and Report," National Council on Competitiveness, http://www.compete.org, 2004, p. 19.

[43] This education team is led by Jeremy Rochelle, Sara Zander, and Charles Patton. The International Organization for Standardization (ISO) is at http://www.iso.org/iso/en/ISOOnline.frontpage.

[44] See "Opening Address at 2004 Teachers' Conference," by Tharman Shanmugaratnam, Ministry of Education, at http://www.moe.gov.sg/speeches/2004/sp20040608.htm.

[45] There is a growing literature on the skills needed in an innovative culture. See "Face Value: The Evangelist of Entrepreneurship," *Economist,* November 5, 2005, p. 72. As Carl Schramm, head of the charitable Ewing Marion Kauffman Foundation, notes, "Entrepreneurs mostly don't come from business schools." The Walt Disney Company and the Kauffman Foundation offer a game for children (see http://www.disney.go.com/hotshot/hsb.html), which teaches them about entrepreneurship.

[46] See, for example, Michael Porter and Scott Stern, "The New Challenge to America's Prosperity: Findings from the Innovation Index," Council on Competitiveness, Washington, D.C., 1999, p.37.

[47] See some specific suggestions in the "National Innovation Agenda," p. 11,

"Introduction to National Innovation Initiative Summit and Report," Council on Competitiveness, Washington, D.C., 2005.

48 "National Innovation Initiative Summit and Report," National Council on Competitiveness, Washington, D.C., 2005, p. 42.

49 See Worcester Polytechnic Institute at http://www.wpi.edu.

50 David Nordfors, "The Concept of Innovation Journalism and a Programme for Developing It," *VINNOVA Information, VI,* May 2003. Also see Innovation Journalism, Vol. 1, No. 1, May 2004, at http://innovationjournalism.org.

51 See "United States," World Fact Book at http://www.cia.gov/cia/publications/factbook/geos/us.html.

52 See the U.S. Department of Labor, Bureau of Labor Statistics at http://www.bls.gov/home.htm.

53 Christine Tierney, "Big 3 Market Share Dips to All-Time Low," *Auto Insider,* January 5, 2005, at http://www.detnews.com.

54 Rhonda Abrams, "All Hat, No Cattle," Inc.com, March 2004, at http://www.inc.com/articles/2004/03/allhatnocattle.html. "During the last decade, small business has become an even more important part of the American economy. Indeed, if it weren't for the jobs created by small business, the American employment picture would be far more dismal than it is. Virtually every study of new job creation shows that it's small and new companies—not big or existing businesses—creating jobs. A 2003 Census Department study found most job growth came from companies less than two years old. 'The small business sector of America's economy is effectively the second largest economy in the world,' according to Donald Wilson, president of the Association of Small Business Development Centers. This exceeds the economies of Germany, France, and Great Britain combined. Our nation's gross domestic product for 2002 was $10.2 trillion . . . the small business sector was just over $5.3 trillion." One interesting side note: "More than half the new jobs created in U.S. since 1992 are attributed to women-owned businesses." See Connie Sitterly, "Women . . . Excel!," Sittcom at http://www.sittcom.com/women/excel.htm.

55 "National Innovation Initiative Summit and Report," National Council on Competitiveness, Washington, D.C., 2005, p. 44.

56 *Ibid.* Consult this source for the original data supporting all these conclusions.

57 Jim Hopkins, "U.S. Entrepreneurial Spirit Remains Steady, Study Shows," *USA Today,* November 28, 2005. The Ewing Marion Kauffman Foundation is doing important work in studying and supporting innovation and entrepreneurship. See more at http://www.kauffman.org/.

58 Betsy Seanard and Lloyd J. Taylor, "Goldratt's Thinking Process Applied to the Problem of Business Failures," ASBE, Albuquerque, NM, March 2004, citing A. Garcia of the SBA.

59 See the Small Business Administration at http://www.sba.gov.

[60] Curiously, at about the same time socialist countries embraced the same basic idea—the central government knows best. The similarities of these systems with those developed by the robber barons are ironic.

[61] For GDP rankings, see "Country Ranks 2005," Countries of the World, at http://www.photius.com/rankings/; and if you want a concise tutorial on GDP, see "GDP," Wikipedia, at http://www.wikipedia.org.

[62] Bernard Wysocki, Jr., "For Economist Paul Romer, Prosperity Depends on Ideas," *Wall Street Journal,* January 21, 1997.

[63] Maslow's hierarchy is available in many sources. For the original go to A. H. Maslow, "A Theory of Human Motivation," *Psychological Review* 50(4), 1943, pp. 370–96.

[64] Disneyland's fiftieth's anniversary was celebrated July 17, 1955.

APPENDIX: VALUE FACTOR ANALYSIS: QUANTIFYING CUSTOMER VALUE

[1] The word *customer* in brackets was added by us to conform to our definition of customer value.

[2] VFA assumes that subjective attributes, such as film quality, can be included with tangible ones, like film resolution. For example, a high-resolution image of a mountain scene can be rated as having the same overall picture quality as a low-resolution image of your child. Additional study is needed but by using VFA you will more quickly identify those attributes that are important to your customers.

[3] QFD (Quality Function Deployment) has many advantages, including helping to identify "critical to quality" issues by using cross-functional teams to represent the "voice of the customer" in the analysis of products, services, and processes. As the QFD Institute at http://www.qfdi.org says, "QFD links the needs of the customer (end user) with design, development, engineering, manufacturing, and service functions. It helps organizations seek out both spoken and unspoken needs, translate these into actions and designs, and focus various business functions toward achieving this common goal. QFD empowers organizations to exceed normal expectations and provide a level of unanticipated excitement that generates value."

acknowledgments

We are indebted to literally hundreds of people who have added ideas, made suggestions, and supported us throughout this endeavor. SRI International and the Sarnoff Corporation provided stimulating work environments filled with brilliant, passionate colleagues where these ideas were forged, reshaped, reheated, and recast again and again. If you could turn back the clock to when Sarnoff first joined SRI, you would find Norman Winarsky and Curt working late into the night, trying to isolate the essential elements of innovation. Norman was a key architect for many of these ideas and deserves special recognition. The seeds of all these ideas began with Norman and others at Sarnoff and continued unabated at SRI.

Within SRI, people from all levels of the organization continued to sharpen the ideas. SRI teammates generously participated in many Watering Hole meetings, both within SRI and with clients during the SRI Discipline of Innovation workshops. Within the workshops, SRI individuals gave their time and insights, especially Larry Dubois, Peter Marcotullio, Bill Mark, Grace Chou, Gary Bridges, and Len Polizzotto.

A special thanks to those at SRI who either read earlier versions of the manuscript or gave invaluable suggestions, including Dennis Beatrice, Norman Winarsky, Bill Mark, John Prausa, Gary Bridges, Jeanie Tooker, Walter Moos, Richard Abramson, Don Nielson, Jon Mirsalis, Pat Lincoln, Philip Von Guggenberg, John Bashkin, Judith Lee, Gary Swan, Scott Seaton, Ellie Javadi, Mary Wagner, Ray Perrault, and Tom Furst. Alice Resnick made many key contributions both conceptually and editorially. Thank you. Doug Engelbart remains a constant source of ideas and inspiration. Also at SRI, we greatly appreciate Val Nielsen, the assistant to Curt Carlson, who performed numerous support activities and, remarkably, could always get Bill and Curt in touch with each other.

At the Sarnoff Corporation, SRI's subsidiary, we owe a special debt to Jim Carnes, former CEO; Glenn Reitmeier of the HDTV team; and Satyam Cherukuri, the current CEO. It was Jim Carnes who first brought Bill to work with the Sarnoff team, which laid the groundwork for multiple visits and the forging of a relationship between Bill and Curt. Later, when the SRI Discipline of Innovation team did our first workshop at Sarnoff, the participants gave incredibly helpful feedback. When Curt left Sarnoff to become CEO of SRI, Bill planted the seed that began our journey together. After Curt explained the NABC template to Bill, he said "Curt, this NABC template is really effective. We should write a book about it." That observation started our journey together, with the book and workshop developing together.

Our colleagues in the SRI Discipline of Innovation workshops at SRI have been inspirational. Herman Gyr has a profound understanding of organizational structure and he remains the ultimate visionary. His work with the BBC and others cycled back and continuously improved our workshops. Laszlo Gyorffy has been the magnificent facilitator for the workshops, as well as continuing to refine our instructional approach and techniques. Len Polizzotto, the originator of Value Factor Analysis and other concepts, organized many of the workshops and taught many of the crucial units. His enthusiasm and positive energy continue to motivate. Scott Bramwell has been our graphics partner for the SRI Discipline of Innovation workshops and he has contributed many splendid ideas.

The Creekside Inn in Palo Alto, where Bill always stayed on his visits to SRI, deserves thanks. John and Jennifer at the front desk, we appreciate your forbearance every time we would rearrange room 217 so we could work from 7 to 7, days on end. The housekeepers could not understand why we didn't want the floor vacuumed or any of the piles of paper disturbed—and *gracias* to them for sneaking in to clean those parts of the room we didn't overwhelm with flip charts.

The road to a quality publisher is tortuous. It was Alayne Reesberg who, in a phone call, suggested we call her friend Helen Rees, an agent. Helen, in turn, got us in touch with James Levine, who worked with us to polish our proposal and add essential ideas. It was Jim's professional links with John Mahaney at Crown Business that opened the door to this fine imprint of Random House. John, thank you for all the terrific suggestions on everything from chapter titles, to an insistence on engaging the reader, to helping us better spell out concepts with clarity. It has been a rewarding pleasure working with you.

There are others who deserve special recognition, including Donna Ebel and family, Eleanor Montagna and family, Jim Scott and family, Jack Gelfand, Roger Cohen, Istvan Gorog, Bill Burke, Carmen Catanese, Robert Bartolini, Mike Ettenberg, Ed Johnson, Maggie Johnson, Alan Herzig, Paul Cook, Deb Dinardo, Gary Morgenthaler, Yogen Dalal, George Cody, Richard Peskin, Peter Burt, Deepham Mishra, Frank Guarnieri, Norm Goldsmith, Gooitzen Van Der Wal, Ann-Marie Lanzillotto, Jim Bergen, Sam Grant, Bill Mullarie, Eric Geiger, Jerome Barnum, Bill Webster, Joe Kovacs, Lisbeth Winarsky, Joan Thomson, and Geoffrey Michaels. Others read the manuscript and/or added important insights, including Peter Carlson, Karl Knop, Peter Seitz, Harry Cook, Aline Johnson, and Steve Schlosstein.

Flora and Raymond Carlson created the foundation and exemplified the values. Barbara Sand and family remain role models and inspirations.

Dudley Brown Carlson and Melanie Trost, our spouses, deserve special kudos and thanks. Dudley was there in the earliest days when the manuscript was first being forged and provided helpful critiques multiple times. Both Dudley and Melanie listened to us talk endlessly and provided great sustenance for the creative thinking needed. You have both been enormously helpful in every way. Thank you.

We are delighted that so many people have positively reacted to these ideas, generously giving of their time and insights, while asking for nothing in return. It is a pleasure to know such quality people and to be able to work with them during this important, exciting time. It is truly a world of abundance.

INDEX

A

abundance, 3, 5, 208, 289–91
achievement, motivation and, 221–23
action plans, 195, 196
adaptation, 17–18, 35–36, 291
alternatives
  to innovation best practices,
    287–89
  innovation plans and, 143
  in NABC model, 87, 88, 89, 93, 96
Amazon.com, 32, 60
aNABC value proposition, 105,
  242–43
Apple, 7, 37, 65, 76, 147–48, 173,
  290
approach
  innovation plans and, 142–43
  in NABC model, 10, 39, 86, 88–91,
    94–95
Armstrong, Edwin, 37
Arthur Andersen, 17
Artificial Muscle, Inc., 2, 111, 146, 148
AT&T, 17
audience needs, 105, 242–43
automobile industry
  exponential economy and, 31, 32
  just in time manufacturing, 243
  mass production, 18, 250, 253
  quality improvement, 18, 248–51,
    286
  U.S. carmaker profitability, 284
  See also specific companies

B

Baker, Rick, 278–79
Baldor, 88
Barnum, Jerome, 189
BBC. *See* British Broadcasting
  Corporation
beachhead markets, 139, 140, 143–44
benefits per costs, 31
  customer value and, 70–71, 73–74,
    76–81
  innovation improvement and,
    284–87
  innovation plans and, 143
  in NABC model, 10, 39, 86–95
Bergen, Jim, 221
Berquist, B. J., 177–78
best practices. *See* innovation best
  practices
BMW, 74
Bootstrap Institute, 179
bootstrapping, 171–72
bottom-up strategies, 252–54
brainstorming, 45, 106–7
British Broadcasting Corporation
  (BBC), 15
  public value and, 69–70, 105, 108
  value proposition for, 105, 242–43
  Watering Holes and, 101–6, 108,
    109, 113, 151–52
Buffett, Warren, 293
Burroughs, 28
Burt, Peter, 220–21, 222, 228

about the authors

CURTIS R. CARLSON

Curt is president and CEO of SRI International in Menlo Park, California, which is in the heart of Silicon Valley. He worked first for the RCA Sarnoff Laboratories in Princeton, N.J., and then for GE before SRI International acquired the Sarnoff Laboratories from GE. Curt started and helped lead the team that, with other companies, set the U.S. standard for digital HDTV and that shared an Emmy® Award in 1997 for broadcasting excellence. His team won a second Emmy in 2000 for additional contributions to broadcasting. Curt has been on many commercial and government boards and was awarded the Otto Schade prize from the Society of Information Display for his contributions to imaging sciences. At the Sarnoff Corporation he was head of new ventures, where he helped form over a dozen new companies. He helped create the SRI Discipline of Innovation workshop, which is given to leading world-wide enterprises.

Curt started his education at Worcester Polytechnic Institute in Worcester, Massachusetts, majoring in physics, and then went on to receive his Ph.D. from Rutgers University, in New Brunswick, N.J. At fifteen he won the Rhode Island All American Soap Box Derby while he was also playing the violin professionally in the Rhode Island Philharmonic Orchestra. Music remains his primary avocation.

WILLIAM W. WILMOT

As director of the Collaboration Institute, Bill works internationally with all types of organizations—corporate, nonprofit, and educational. His specialty is helping others move past disputes and into productive relationships. He has worked with more than 300 enterprises and is recognized as an Advanced Practitioner in the Association for Conflict Resolution. He has written six other books, including the coauthored books *Interpersonal Conflict,* seventh edition; and *Artful Mediation: Constructive Conflict at Work.* He is also a senior associate with the Yarbrough Group, where he and Elaine Yarbrough focus on organizational development and conflict work. Bill helped create the SRI Discipline of Innovation workshop, where he is a partner and instructor.

Bill is a professor emeritus at the University of Montana, where he is still active as a consultant and trainer. His Ph.D. is from the University of Washington. He is a dedicated outdoors enthusiast whose adventures have taken him from his home base in the Rocky Mountains to the Himalayas for treks in Nepal, Bhutan, and Tibet.